Alienation and Value-Neutrality

A. J. LOUGHLIN

Routledge

Taylor & Francis Group

LONDON AND NEW YORK

First published 1998 by Ashgate Publishing

Reissued 2018 by Routledge
2 Park Square, Milton Park, Abingdon, Oxon, OX14 4RN
711 Third Avenue, New York, NY I 0017, USA

Routledge is an imprint of the Taylor & Francis Group, an informa business

Publisher's Note
The publisher has gone to great lengths to ensure the quality of this reprint but points out that some imperfections in the original copies may be apparent.

Disclaimer
The publisher has made every effort to trace copyright holders and welcomes correspondence from those they have been unable to contact.

A Library of Congress record exists under LC control number: 98047213

ISBN 13: 978-1-138-60810-8 (hbk)
ISBN 13: 978-1-138-60811-5 (pbk)
ISBN 13: 978-0-429-45948-1 (ebk)

Contents

1 Destructive Conceptions of Rationality

1. Under-estimating the Rational Subject

This book is about a conception of rationality which is so pervasive that its profound influence is scarcely recognised. It determines the way we think about ourselves and our world, and what we mean by 'knowledge', 'progress', 'science' and 'democracy'. Even those thinkers who, believing themselves to be saying something radical, reject or attempt to 'problematise' such notions as objectivity and rationality, are in fact thinking in terms dictated by the concept of reason whose logic I want to expose and reject. I will argue that this conception is *alienating*, meaning that it sets us apart, in a destructive way, from the world we know, and that it does so by causing us to ignore or devalue those subjective capacities which could enrich our awareness of that world. Thus it impoverishes both the rational subject and the world which the subject seeks to comprehend.

Its influence is by no means restricted to the fields of academic philosophy and science. The destructive concept of reason permeates individual consciousness and society at every level. It has profoundly affected the development of the modern, liberal democratic state. It is at the heart of the various, platitudinous expressions of moral relativism that are common place in contemporary societies, and it underlies the view, still popular in many quarters, that only value-neutral science can provide true knowledge. It *frames the debate* in all these areas, passing largely unnoticed and unexamined in the process. Even where it is subjected to criticism, the critics themselves will often seem not to have escaped its pervasive influence. In this work I attempt to explain this conception of reasoning, the effects it has had on the development of natural and social science, and its implications for our thinking about morals and politics.

Early twentieth century thought was dominated by a conception of rationality which sparked a re-evaluation of the nature and status of many traditional subjects. Even the traditional preserves of the philosopher – metaphysics and epistemology – were dismissed as consisting only of meaningless, irresolvable questions by strict empiricists seeking to sweep away all 'nonsense' from what they perceived to be the House of Reason. The conception of rationality at work during this period might well be labelled the 'logical positivist' (LP) conception and it consists in the restriction of what we can rationally claim to know to what can be derived from the empirically ascertainable data, including statements or theories founded on such data. Statements which *are* empirically verifiable provide the raw materials for science. Statements which are not verifiable are no more than sheer nonsense. Since the validity of the inductive principle itself cannot be established either through sense experience or by logic, the only reliable form of reasoning acknowledged by the strict empiricists was *deductive* reasoning. Thus, philosophers and scientists influenced by the ideas of David Hume concentrated on developing accounts of science (such as the Deductive-Nomological model, to be discussed below) which show that science does not depend essentially on what they deemed to be rationally indefensible *inductive* reasoning.

It is perhaps no surprise to us that logical positivists did not get very far in lending support to our ordinary common-sense statements about the world in which we live. Indeed, it must be clear that this was not their aim, for the Vienna Circle[1] did not take its lead from the ordinary individual but, on the contrary, sought to take the lead itself in offering what amounted to a manifesto for setting to rights a somewhat fractious and confused post-war world. Reason, as they conceived it, was to be the driving force behind empirical science, whilst science itself was to provide the material means for progress. Thus, whilst science was elevated in standing, many of our most common-place convictions were undermined. The logical positivist conception of rationality would not sanction the idea that reason could determine such things as right and wrong, that it could be employed to help distinguish a good way to live from a bad way, or help demonstrate God's existence. Even the most common-place assumption that material objects are really 'out there', independent of our will and perceptions, turns out neither to be justified by the senses nor evident to reason. What this particular strain of positivism did was to undermine many of the layperson's ordinary knowledge claims, denying that much of what many people tend to believe actually belongs to the class of *rational* (even *meaningful*) claims. In short, they sought to limit the contents of our world by casting out all beliefs that did not conform to one particular conception of the 'rationally justifiable'.

Now, obviously, if it is the case that what we ordinarily believe is false, or unjustified, then in disabusing us of these errors the logical positivists would clearly be doing humanity a service. It is not enough to say 'we do, as a matter of fact, believe this' or 'we want to believe that', thereby dismissing all those who criticise our rationality. Equally, however, we can ask those who are doing the criticising to tell us what their conception of rationality *does* allow us to believe: what alternative picture of the world are they presenting us with and is it coherent? If a particular conception of what is rational rules out the possibility of rendering our world coherent using reason then we can be forgiven for not taking seriously any criticisms founded on that conception. We might opt, instead, to revise our understanding of the concept of rationality itself. This would represent the beginnings of a legitimate response to the criticisms offered by logical positivists of some of our most cherished beliefs.

What I wish to suggest in what follows is that if we continually underestimate what the rational subject as such can achieve we can render the task of making our world cohere an impossible one. This is essentially what the logical positivists have done and, with regard to developments in the philosophy of science, the long-term results are becoming evident. The rationality of science itself has been questioned (it cannot meet its own stringent criteria; it cannot eliminate from its own house what it has deemed 'non-rational') and, in so far as the world has coherence, it is often thought to be only at the expense of its independence and uniqueness[2]. Given such a *limiting* view of rationality, the subject has been unable to leap the void which, it is imagined, separates it from an independent though ultimately knowable world – and thus the subject concludes that there can *be* no such world. This is a daring conclusion, but perhaps we should explore other, more generous accounts of the nature of rationality before we embrace what is possibly (if there really *is* a world out there) an alienating conclusion.

If we are going to define rationality, rather than simply give up on it, then we have to define it in such a way that there is at least some thing which is an instance of it; at least some beliefs must be capable of being rationally justified, and at least some action, in some respect, must be rationally defensible. If it were not so, the term would become vacuous and redundant.

What I aim to show is that if we give an account of rationality which is overly limiting then ultimately there will be no genuine, non-arbitrarily selected instances of rationally-held belief or of rationally justifiable action. Rationality has to be defined adequately to counter irrationalism and scepticism or else there is little point defining it at all. An adequate definition is, as I will demonstrate (Chapter 3), what the logical positivists failed to provide, but they were by no means alone in adopting an overly limiting conception of rationality. This motif of under-estimating the powers of the rational subject runs throughout the history of philosophy, with logical positivism providing only a more recent instance.

In W.A Weisskopf's book, *Alienation and Economics*[3], the author takes up this theme of the gradual undermining of the concept of rationality and, as the book's title indicates, makes the link between contemporary human alienation and the limiting conception of rationality which he argues is apparent in positivist economic science. Weisskopf identifies the cause of present-day alienation with a pattern of thinking which is characteristic of the scientific, technological age. He calls this mode of reasoning 'means-end', 'technical' or 'instrumental' rationality and links it with a desire, motivating positivist philosophy of all types, to set science on firm philosophical foundations. Whatever cannot be reduced to, or derived from, the data of sensory experience does not belong to science, although there has been much disagreement within the positivist tradition about what sort of things this includes (logical positivists being by far the most stringent). The one thing agreed upon, however, is that the rationality of science requires the elimination of value judgments in order to maintain the claimed objectivity of research. This is where the 'means-end' tag associated with science-orientated conceptions of rationality comes into play. The 'means' can be determined in a value-free (and therefore rational) manner, whilst the 'ends' clearly cannot be. Rationality is limited to the sphere of determining fact and it thus becomes imperative to show that the facts, suggesting appropriate means, are indeed separable from the ends determined by non-rational valuations.

The role of Hume in shaping the 'technical' or 'means-end' conception of rationality which Weisskopf describes is evident. Hume tells us that we cannot logically derive an 'ought' from an 'is'. In other words, we cannot deduce from any number of purely factual statements, another statement which *evaluates* the state of affairs. This is because values, as expressed in moral judgments, are not derived from statements of fact, but rather they depend on a person's *attitude* towards the stated facts. By implication, values are determined by the subject, facts are determined by the object, and a rationally defensible study of the facts will therefore exclude all subjective value judgments or else risk compromising the enquiry's objectivity. The logical positivists' conception of rationality described above is a species of 'means-end' or 'technical' rationality, though it is the one which takes Hume's rigorous empiricism most seriously. Other adherents to the 'means-end' conception do not necessarily share the logical positivists' profound mistrust of induction, though they all agree on the importance of value-neutrality in whatever passes as truly objective science.

Weisskopf suggests that the 'means-end' distinction which permeates both inductivist and deductivist traditions in science shapes not only the way we think about science, but the way we think about rationality as such. In addition, positivist science has determined the way we think about knowledge and our relationship to the world in which we use knowledge to govern action. The 'means-end' mode of thinking casts true knowledge in the light of that which helps us manipulate the world so as to achieve our arbitrarily selected goals.

The formulation of the 'Deductive-Nomological' (DN) model of scientific explanation illustrates well the manner in which we have come to equate knowledge with that which enables us to predict and manipulate the environment. The DN model equates the power of explanation with the possession of theoretical tools for yielding successful predictions such that, if a thing *can* be explained, it can be anticipated also. However, the predominance of this view of explanation has had far-reaching effects on other modes of explaining, or understanding, the world. Non-scientific modes of explanation have suffered a lowering of status in the public's eyes, and even the *social* sciences, which cannot always adequately predict and yield tools for control, are deemed somehow inferior. In contrast, the natural, or 'pure', sciences have risen high in people's estimation and even today people still look to science to solve humanity's problems. Thus, the science-dominated view of knowledge and of rational enquiry has usurped the place of a fuller conception of rationality, one which would admit of the possibility of other modes of rational explanation and perhaps even suggest other ways of interacting with the world than that of driving it into the service of our apparently arbitrary whims. Weisskopf gives a name to this 'fuller conception': he calls it 'encompassing' or 'ontological' rationality, but we will return to an examination of this concept later (Chapters 2 and 3).

First, I want to suggest that a limiting, and indeed alienating, conception of rationality is not unique to the many variants of positivism. The ultimate source of human alienation spans both empiricist *and* rationalist philosophical traditions.

The Age of Science is often said to have its origins, or at least to have drawn support in its infancy, from the writings of Descartes. Like the positivists, Descartes' concern is to provide a sound methodology and firm philosophical foundations for the study of science, but his methodological scepticism betrays the fact that he too possesses only a limiting conception of rationality. He systematically questions all of our ordinary knowledge assumptions: the belief in external objects, in other minds, our belief that we can distinguish between illusion and reality, and finally, the conviction about the existence of our own bodies. Having thus removed all the contents from our world, he seeks to put everything back again, only this time justifying our common-sensical beliefs by reference to a rationalist proof of the existence of God and the argument that God cannot be a deceiver. Descartes tries to deduce from the supposedly indubitable premise that 'I exist and have a conception of the Infinite Being' that he cannot be deceived most of the time and is therefore justified in believing most of the things which he believes ordinarily.

Unfortunately, Descartes' argument fails to reassure us because inherent in this approach is a view of rationality which can only serve to undermine many of our knowledge claims. In other words, he does not concede to the rational subject itself sufficient powers to repel scepticism. It is interesting to note however that ultimately Descartes has to appeal to God to provide the basis for refurbishing his world. His ontological argument (even if it were not itself open to criticism) fails to convince anyone who is not already a believer to believe in God. The ontological argument cannot instil faith and I suspect that Descartes himself was not much moved by it. (His faith doubtless preceded the formulation of the argument.) Thus, at the foundations of his attempt to justify ordinary knowledge claims we find a simple act of faith. This is not to say that his belief is therefore irrational. Quite the contrary. Assuming that Descartes is right and God does exist, then his act of faith represents the engagement of a rational subject with something beyond itself. It is an engagement of the *whole* self (a subject with a particular history and experiences, attempting to use its various capacities, including its imagination and creativity, to make sense of its life) and not merely an exercise in deductive logic. Real faith (as opposed to its mere expression in words) is likely to be inspired by some features of the subject's experience, such that the belief in God *makes sense* within the context of that experience. In the same way, we have confidence in the existence of external objects and of other people because we experience them as an integral part of our lives and because our experiences would not make sense if we did *not* believe in such things. It is only this kind of experience which can inspire confidence. No purely logical argument can do so because such arguments do not engage the whole of the rational self.

I refer to Descartes at this point because I believe it shows how a limiting conception of rationality can lead to a dismantling of what we would wish to claim as our established body of knowledge. Once we set off down that road, as it seems Descartes does, we may find that we cannot get back again – or, at least, not without some sleight of hand[4]. I do not wish to suggest, however, that we should refuse to go down the sceptic's road simply as a matter of faith. Rather, I will argue that our faith in the external world *is* a form of knowledge, perfectly rational and legitimate, provided that we take a broader view of logical argument and of the nature of rationality itself.

It strikes me as a peculiarity within the Cartesian tradition that it seems to imply the possibility of understanding the world in a way untainted by the presence of the knower. It is imagined that if knowledge is from a distinctly individual perspective it is merely subjective in nature and therefore not really about anything 'out there' at all. The subject and object are isolated from one another in Descartes' metaphysic: the study of the object is in the sphere of the sciences and is to be kept quite separate from the study of the subject (the self) which, if it is to be known *as subject* is accessible only to the faculty of introspection. Subject and object never merge. They co-exist, but are distinct and interact only in a relatively unsophisticated manner: objects seem to impose certain perceptions upon us, like their texture, solidity and weight, and these perceptions come to us unregulated by our uniquely individual pre-dispositions to perceive them. It follows that the Cartesian conception of science is an impersonal, mechanical one, granting no role to individual creativity. Scientists merely look at the world in a detached manner and wait for objects to reveal their nature to them. Thus, inspite of the radically different starting point, Descartes' conception of science shares in common with the conception shaped by British Empiricism, the assumption that personal value judgments play no part in good science. The view dominant throughout scientific history is essentially this: the attainment of objective scientific knowledge requires the passivity, or emotional detachment, of the truly rational enquiring subject.

The problem with this particular conception of understanding is that, because it allows for no fuller interaction of the subject with the object of interest (the world), it becomes all too easy to cast doubt on the very existence of the external world. This, as we know, is what Descartes does and it leads him to systematically remove all objects from the sphere of the indubitable, leaving behind only the thinking subject at the mercy of the Evil Daemon. The contents of the world, and our reasons for action and interaction with the contents of that world, cannot be restored on the Cartesian conception of rationality because it has already built into it the denial of our ability to know the world directly, to establish the reality of objects for ourselves through *actively* bringing them into our awareness.

To know something, I wish to suggest, is to act in a particular way such that if we insist on the scientist being passive (in the sense that the conventional understanding of 'objectivity' requires) then the task of bridging the gap between the subject's 'in here' and the world's 'out there' becomes impossible. Effectively, a problem has been generated by (1) supposing the isolation of subject from object as a starting point and then (2) handicapping the subject in its attempt to leap the subject/object gap by insisting that it keep one foot on the ground. Can the enquiring scientist really pacify certain features of his/her personality and belief system, whilst at the same time vigorously exercising others, namely those features which are not believed to impair scientific objectivity? Is this a realistic view of how the search for greater knowledge must proceed? Obviously these are the questions which must now be considered. If it transpires that we cannot separate out the different elements of our belief systems, separating the factual claims from value judgments, we will be left with only two options: giving up on the possibility of rationality, or reviewing what we mean by rationality such that it no longer requires the seemingly impossible task of setting aside key elements of one's belief system. The latter half of this chapter will be concerned with examining the plausibility of sustaining a means/end distinction in science, whilst Chapter 6 will go on to examine in greater depth a more sophisticated variant of what is essentially the *same* model of rational, scientific enquiry.

Intellectual detachment seems to be the key to understanding traditional conceptions of scientific objectivity. Whilst it is accepted that scientists must bring enthusiasm to their work, their enthusiasm must not be for any particular explanation over others, except insofar as the actual facts justify such enthusiasm. In other words, the devotion to a theory must not precede the facts which point to that particular theory's being superior to others. This suggests a view of rational knowledge acquisition which places the power of determining the subject's beliefs firmly with the object *right up to the point where those beliefs take the form of a sophisticated, and perhaps even highly abstract, theoretical structure.* It is between such complex structures that the scientist must choose so that, if we wish to defend traditional conceptions of objectivity, we must show that the object (the world we experience) *gives* to the subject sufficient grounds to construct and select theories without the subject ever needing to over-reach itself and violate its own demand for intellectual detachment.

Explaining how theoretical structures can come about objectively (where this means without the incursion of value judgments) is evidently going to be a more difficult task than explaining how individual factual statements can be justified. Scientists cannot wait for the facts to present themselves in such a way as to suggest ready-made theories. They are compelled by the unwillingness of nature to yield up its own self-descriptions and explanations to find some explanation for themselves and then to return to nature only to see if their 'man-made' explanations are a good fit. In the process, they will have to exercise their imagination in interpreting evidence, agree on descriptions of the data, formulate hypotheses, and ultimately test them. Even then, they will have to exercise their judgment to determine what constitutes a 'successful' result because, again, nature will not mark their theoretical successes with a helpful pat on the back.

In practice therefore, I will be arguing that the scientist, like the most undisciplined of laymen condemned by Descartes, must allow him/herself to be guided by the imagination, perhaps also at some level by aesthetic and even moral considerations, or by analogies with personal experience (like Archemedes in the bath). S/he will do so simply because the reticence of nature gives her no other choice; if the object falls short of giving the subject the knowledge sought after then the subject *must* reach out further to get it. The imagination is the very faculty we are endowed with to perform this 'reaching out' operation and the scientist, in using this faculty, would not necessarily be misguided. What we must look for to fill out an alternative account of scientific objectivity, are not further guidelines for achieving greater detachment of the subject from the object, but new guidelines for encouraging a more productive and imaginative *involvement* with the objective world.

The scientist, of course, may eventually be proven wrong in his/her commitment to a particular theory, but I shall be arguing that this possibility arises not because a theory influenced by the imagination (by *subjective* factors) is, in principle, mistaken but because of the possibility that the scientist's perspective on reality, conditioned by the totality of his/her belief system, may have been flawed. The assumption that all respectable scientific theories are at least value-neutral, even if factually wrong, rules out the option of tracing an error back to certain, value-related components of the scientist's belief system. The underlying assumption (referred to above) is that the object *gives* enough of itself, and of its relation to other objects, to make any adjustment of the subject's perspective strictly unnecessary. To acknowledge the need to examine and alter one's perspective in order to attain greater knowledge is to contradict the assumption that objectivity requires the subject to be passive.

The Cartesian view of rationality removes the subject, the perceiver, from the world-picture in the sense that, in order to really know the world, the subject must be rendered transparent. It is a view which suggests that the active striving of the subject towards the very things it wishes to know only prevents it from attaining the sought-after knowledge. And yet Descartes himself suggests that we can intelligibly doubt the existence and nature of all the things we would like to take for granted. He effectively removes the objects of the world from the subject's immediate and private world of experience and leaves them at a distance where the perceiver *thinks* s/he can see them but cannot (at least without God's help) reach out and touch them, to confirm their reality and banish doubts. Without confidence in one's perceptions, in the reality of the world in which one lives, the Cartesian self undermines not only the knowledge which is there for the taking, but with that, loses all reason for action. An impersonal, mechanistic world offers no suggestions as to what should be our proper goals and guiding principles. This aspect of Cartesian philosophy brings to our attention a further way in which we can be said to be alienated.

Science (at least prior to the Quantum Physics revolution) has come to be based on the limiting, Cartesian conception of rationality described above. The physical world has been viewed as a predictable and controllable mechanism indifferent to, and independent of, the human mind. However, whereas this mechanistic world picture has lost its grip on the minds of many twentieth century physicists, it still (as Philip Mirowski points out in *Against Mechanism*[5]) keeps a tight hold on the minds of social scientists. In mainstream economic theory, for instance, Mirowski suggests that the tendency to view individual economic agents as insignificant (that is, not as individuals at all) is an expression of the underlying, impersonal, mechanistic world view. All economic agents are judged to function in accordance with one universally applicable principle of rationality in much the same way as all material objects fall to the ground in accordance with the law of gravity. Just as Descartes sees the role of philosophy as one of defining the rules of scientific method, providing rules for procedure which could resolve any dispute and guarantee scientific validity to the victor, so the modern economist is concerned with defining the rules of market organisation and the principles underlying the behaviour of economic agents. The market, like the natural world, will then automatically and impartially operate to distribute resources in a way which appears legitimate by virtue of its mechanical indifference.[6]

The Cartesian tradition in science opens up a gulf between the history of science and science itself because the latter is concerned with the independent, objective world and the former, with the development of ideas, that is, with the workings of the imagination of individual scientists and the interchange of their ideas. In the same way Mirowski traces the influence of the Cartesian tradition of thought on neo-classical economics. He notes the marked separation of the study of economics from the study of the histories of particular economies.

Most economists seem to imagine there is little to be yielded from a study of people's ways of thinking about themselves, of their ideologies and the beliefs which will, in some way, influence how they react and, in turn, will be acted upon by their environment. Weisskopf also acknowledges the complex interaction process between individuals and society. In *Alienation and Economics* he observes that economics is an expression of the prevailing value-attitudes or belief system. Economic reality, and economic theory itself, determine how people perceive things and are influenced in turn by people's perceptions of them. Economic theory is shaped by the attitudes of the time yet also 'feeds back' into society, reinforcing these attitudes, carrying them through to the next stage of development or even stimulating a reaction against them. The point which critical social theorists have made[7] is that since value-attitudes undoubtedly do shape the theories of allegedly value-neutral social scientists, it is as well to identify what these attitudes

11

are, to assess the underlying belief system, and to see if their ideas are acceptable. Since the ideas of the neo-classical school of economics spring from a 'technical' conception of rationality, a conception whose dominance owes as much to the Cartesian tradition in science as it does to the empiricist tradition, we find that the belief system of the modern economist *per se* is one devoid of values. Not surprisingly, the social scientist can find no suggestion of appropriate values in a cold, mechanistic world and, as a result, theories which are designed to explain this unfamiliar world are not merely value-neutral, they are also, in Weisskopf's opinion, 'value-empty' because they are underpinned by a conception of rationality which denies the objectivity of values.

In a world taken apart by the attempted application of an impoverished conception of rationality, a world where we can have no confidence in the permanence or reality of objects and other people, we find no other principle of 'rational' action but the economist's principle of rational self-interest (though in fact it is not even clear why we should be *self*-interested). This is the only principle the economist can offer to spur the producer/consumer into action.

For Weisskopf, this motivation is inadequate. He takes a broader view of the nature of human rationality and argues that we are failing to live up to our potential when only one aspect of our rational faculties is acknowledged. We are, he says, 'reduced to a part of what we could be'. In short, we are alienated.

The origins of what Weisskopf refers to as 'the disintegration of rationality' are diverse, but we see traces of an alienating way of conceiving of the rational subject present throughout philosophical history. We can, for instance, trace the present-day tendency to bestow the highest intellectual status on science, whilst demoting or even rejecting the validity of other types of discourse, to Kant's division of rationality into Pure and Practical Reason. This purely clarificatory device conveys the impression that we possess two quite distinct faculties of reason, capable of operating independently of one another. Taken together, pure and practical reason may well be seen as roughly equivalent to Weisskopf's 'encompassing' or 'ontological' rationality (about which more will be said presently). However, since Kant, the splitting off of pure from practical reason has weakened the status of the latter, exposing Kant's moral philosophy to criticism and finally resulting in the rejection of the possibility of objective moral truths. This permits sceptics in general to exercise their scepticism against objectivity in morals but at the same time to preserve their confidence in science, in inductive reasoning, and in the possibility of the value-neutral study of the physical world. Thus Kant's successors (among them neo-Kantians who were known to have influenced Weber) have played a crucial role in divorcing facts from values and in promoting the positivist doctrine of value-freedom in natural and social science.

Richard Kroner, one of the neo-Kantians of the Heidleberg school, remarks on the basis of his interpretation of Kant that

> the world in which we as moral beings act and pursue our ends obviously cannot be penetrated by mathematical knowledge; therefore this world cannot be grasped in its reality by any theoretical [ie. scientific] means.

Kroner's statement (from *Kant's Weltanschauung*) clearly expresses the mistaken notion that the world in which we act as moral beings possesses a reality distinct from the world in which we perceive and comprehend natural phenomena. But there *is* only one reality and we perceive and act within this one world. Kant's noumenal/phenomenal world distinction divides the world into two halves, one with a somewhat mystical, suprasensible status, and in so doing he gives ground to the more limited conception of what constitutes 'the real world'.

Having discarded the idea of a 'noumenal' world, the neo-Kantians found themselves pushed further towards total idealism, ultimately collapsing the knower/known distinction into the realm of the knower or observer. From there it is but a short step to the idea that it is the subject who actually bestows on the world its coherence: reality is what it is because we make it so. Thus, since Kant, the emphasis in philosophy has shifted away from the idea of an objective, independent world towards a greater interest in the subject's ways of *speaking* about the world. The main emphasis has been on the philosophy of language, leading to a number of distortions and a departure from what the philosophical layperson would tend to see as common-sense. The philosophy of language represents an approach to philosophy which has its foundation in the limiting, destructive conception of rationality I have been discussing.

One of the first victims of such a conception has been the belief in the objectivity of morals, and although many accepted the demolition of one of the most important aspects of ordinary knowledge, there were others who fought (albeit within the confines of the linguistic tradition) to salvage something of objectivist moral philosophy in some new form. Hare is one such philosopher (discussed in Chapter 5). There are also attempts by Apel[8], Habermas[9] and Popper[10], though all suffer from the fact that they never explicitly identify and reject the destructive, subject-limiting rationality of modern Western thought.

Habermas tries to argue for the possibility of rational moral discourse by showing that the possibility of agreement on objective truths is a presupposition of our use of language. With Popper, it is assumed that an initial 'leap of faith' is necessary before we are bound by the presuppositions of rational discourse, though the leap of faith itself must remain an irrational act. This conclusion has displeasing consequences for Popper's own philosophy since it ultimately renders him powerless to refute the determined sceptic (see Chapter 3). Others, like Apel, try to argue that Popper is wrong not to extend the concept of rationality to the presuppositions of language itself such that no language-user could practically avoid being bound by the constraints of rational discourse. None of us can avoid involvement in the community, thus avoiding the use of language. Consequently, we are all irrevocably committed to rational discourse and to the belief in the possibility of universal agreement and objective truths.

But this defence of objectivism will not do. It appears to confuse the descriptively correct assertion that human beings are, as a matter of fact, sociable and communicative with the quite different evaluative assertion that human beings *ought* to communicate in accordance with (Apel's) rational presuppositions. The point, presumably, which sceptics would wish to make is that nothing compels us, even as compulsive language-users, to act in accordance with Apel's presuppositions which are so tailored as to allow for the possibility of agreement and objective knowledge. It will not suffice to say that because successful communication demands a mutual commitment to (for instance) truth and consistency, we are all thereby bound by these communicative ideals, for what compels any of us to value the goal of *successful* communication or to interpret this goal in the same terms?

The shortcomings of this mode of argument can be traced to the fact that its exponents do not explicitly extend the concept of rationality beyond language-use, beyond even the *presuppositions* of language-use, to the possession of consciousness itself. It is not that the possession of consciousness is, on its own, sufficient for an entity to be deemed rational (for it is obviously not) but rather that it is at least minimally necessary for

14

the *possibility* of assessing something as 'rational' behaviour. Without some kind of brute awareness of the context for action, nothing could qualify as a rational response: there would be nothing that the 'action' could be seen as a response to. Consciousness is the means by which reality impinges upon us and as such it is what opens up the possibility of judging behaviour as rational or irrational with respect to the world which consciousness brings into view. It is because we are conscious beings, aware of the context for action, that we cannot escape the respective labels of 'rational' or 'irrational'. If we are bound by rational requirements in virtue of the very fact of our consciousness, then declining to accept the presuppositions of language-use will serve as no excuse for even if we could retreat from the discursive community, we could retreat no further without passing out of existence altogether. In other words, irrationalism is a tenable position for no-one.

If the application of the concept of rationality is not extended beyond language-use, if its possibility is not linked to consciousness itself, then there will always remain a gap into which we may insert the question: 'Why *should* we accept the conditions of rational discourse?' Under the influence of the philosophy of language, philosophers have been disinclined to see language as merely an expression (though perhaps an imperfect one) of our direct involvement with an independent, objective world, an involvement which we cannot choose to deny or reject without incurring legitimate criticism for irrationality. We are already conscious beings acting in, and interacting with, this world and the nature of the world will not change simply because we choose to deny it or re-describe it.

I am proposing here that rationality be seen in the broadest terms as that which assists conscious beings in their activities, negotiating a complex and sometimes hostile world. Language-use may assist in this process, but it could equally hinder us if we mistake the constructs of language for reality itself. Rationality (or, to adopt Weisskopf's terminology, 'encompassing' rationality) is what enables us to bring more of the real world into the light of conscious awareness; the further we are able to extend our awareness and allow it to inform action, the greater is our rational potential. The exercise of encompassing rationality implies a willingness to try to embrace all that the world really is, whilst to be alienated is to push away aspects of reality. An alienated subject is isolated through its own irrational choices from features of the world which could otherwise be known.

2. Means-end Rationality

A fuller conception of rationality, such as is being considered here, requires the possibility of making an alienated choice, a choice which sets the subject at a distance from some part of reality or from some (or even all) people. It is a view of rationality which implies that there are good reasons for choosing in a certain way, that rationality extends to the sphere of choosing our principles of action, and that it provides the grounds for selection of ultimate goals. As such it may be characterised as a *constructive* conception of rationality. It promises not only to restore our confidence in the objective world, but to bring us into a fuller interaction with that world and to give us non-arbitrary reasons for action.

Let us suppose for a moment that preferences, value-ordering, and the selection of ultimate goals are not amenable to rationality, that rationality itself consists only in pursuing our arbitrarily chosen ends with the utmost efficiency. In other words, rationality is concerned with comprehending the world so that we may control and manipulate it and thus achieve our given ends. We may then ask: where does this requirement of efficiency come from? Is this a rational requirement or is it merely a particular whim that our ends are achieved in a way characterised by efficiency? Might we not equally choose to pursue our ends in a completely erratic, non-sensical way that wastes the maximum of time and resources whilst still, eventually, achieving our goal? Of course, this seems intuitively ridiculous but the reason it is so is because we all do, as a matter of fact, value time, resources and effort. This happens to be an evaluation that all natural and social scientists agree upon, although they might add that the requirement of efficiency is rational because if our actions were not characterised by efficiency (but non-sensical and erratic as I have described) then we would probably not have survived as a species. What is conducive to survival might plausibly be claimed to be definitive of rationality. They might further add that conceptually linked with the notion of an 'end' is the idea that it is what we would wish to achieve most quickly, that everything set between us and our goal is a mere hindrance to be got out of the way with the minimum of fuss. In other words, with 'ends' being the kind of thing they are, we would naturally seek to minimise the distance between us and them (ie. our attainment of them) and this is best achieved by choosing the most efficient means.

16

I would not wish to deny any of this. Rather, the point I would wish to make is that in employing the concept of efficiency we are unavoidably engaged in making value judgments. The value judgment (which in this case is the judgment that I/the species to which I belong wish(es) to survive) cannot be separated from the action necessary to attain this end. Choosing survival entails a value judgment as can be seen by considering its converse – suicide – which most certainly involves an evaluation of the worth/worthlessness of life. As such, the concept of 'technical' or 'means-end' rationality is shown to be an impure one, that is, it is a concept which draws upon a broader conception of rationality to give content to the requirement of efficiency. No-one can *purely*, so to speak, exercise their faculty of means-end rationality because to select means commits them to some course of action, and in all action there is the possibility of having to evaluate alternatives, of having to choose *between* alternatives.

The requirement of efficiency introduces some very problematic elements into any attempt to determine appropriate means and it confuses the issue of what is 'rational' by means-end criteria. Having settled upon a goal and in casting around for methods of achieving it, we are faced with questions about what means may be classed as 'efficient'. It is not so simple as merely ascertaining the fastest, easiest route to obtaining object X because what may appear to be the fastest route may also involve action with effects which contravene other non-rationally determined goals. For instance, I may be best able to develop my skills as a violinist by joining an amateur orchestra but I find that this particular course of action, though undoubtedly effective, conflicts with my other goals, including completing this book and spending time with my daughter. Consequently, my revised view of an efficient solution in this case involves simply practising at home whenever I can.

This sort of conflict of goals is not unusual when considering what are appropriate means. Indeed, theoretically, every stage in the sequence of actions which constitute the means to attaining an end could throw up new conflicts between competing goals. It is not the case that we can decide on our ultimate goals and then simply let our faculty of means-end rationality go to work to discover the most efficient ways of achieving these goals. What is admissible as an efficient solution is regulated by evaluative criteria.

Russell Keat in *The Politics of Social Theory*[11] acknowledges the complexity of the relationship between means and ends but nevertheless wishes to maintain that we can pursue a value-neutral social science. Such a

17

belief (unfortunately a common one among social theorists) is based upon the assumption that we can, even to some extent, allow means-end rationality to go to work on its own, momentarily forgetting evaluative criteria, or at least leaving the business of evaluation and goal selection to politicians and others. But to think that means-end rationality exists at all independently of value-judgments is a mistake because the concept of means to an end implies action, and action – because it always has an outcome – is never value-neutral. It is always value-regulated.

If we allow ourselves to imagine that means-end rationality is a faculty in its own right, and that it can guide our behaviour towards the rational attainment of goals, we will be disappointed to find that it in fact destroys the very possibility of rationally defending any action whatsoever. This is because it is always open to us to deny the so-called 'efficiency' of its recommendations. Means-end rationality must be seen as an aspect of our rational faculties which can only be distinguished intellectually but never practically isolated from rationality as a whole (where this is understood to be something like Weisskopf's encompassing rationality). Means-end rationality itself cannot be specifically defined without filling out the concept of efficiency and in so doing we find 'rationality' to be something more than it purports.

A conception of rationality which cannot justify or compel any course of action – even as a 'means' to something else – is no form of rationality at all and only passes as such because of its concealed reliance on a broader conception which helps make sense of the notion of 'efficient means'. In so far as the essence of life is activity, means-end rationality reduces our existence to absurdity because *on its own* it offers no constructive principles for action.[12]

One might argue that within the confines of natural science it is possible to obtain a wide basis of agreement upon the appropriate methods for achieving particular results. If a scientist wishes to produce a quantity of copper sulphate solution s/he knows that the appropriate method will be to combine copper oxide with sulphuric acid. Surely no one could dispute that here we have an example of pure technical ('means-end') rationality in operation? Where many people might be prepared to concede that *social* science is inextricably bound up with value-judgments (because of the problems of filling out the concept of efficiency in a domain where every means-act has social ramifications) relatively few would accept that the same difficulties plague the natural scientist. I would certainly agree that there are substantial differences between the two areas of enquiry and that the problem of achieving anything like the appearance of value-neutrality in social science is very much greater. However, in general, I think the dissimilarity of the social to the natural sciences has been over-emphasised. Let us go back to the apparently straightforward example above. The means seems wholly uncontroversial and admissible as an efficient solution to the problem of attaining copper sulphate whatever may be the other aspects of one's belief system. In other words, there seems to be no valuation of any kind involved in reaching the conclusion that combining copper oxide with sulphuric acid is the best, indeed the only way of achieving the desired result. The means to this particular end is simply a matter of fact, a fact which science has uncovered without reference to value-judgments of any kind.

What is correct here is that if the means described really is the only way to produce the relevant substance then the statement of this chemical relationship is, purely and simply, the statement of a fact. This fact of nature would not be otherwise no matter what we thought about it. However, the question arises: how do we *know* it to be a fact? Certainly, the fact is what it is regardless of valuations (and in this sense facts in natural science are obviously separable from values) but our *knowledge* of the fact may depend on what value-determined stance we adopt towards it. If we approach nature from the wrong angle, so to speak, we may be unable to apprehend what others see as obvious. Thus, it was not always within the grasp of earlier scientists to recognise even the more elementary facts of modern chemistry implicit within the example above. Certain theoretical developments had to be in place before scientists could move towards a better understanding of molecular structure and of the processes of oxidation and reduction which lie behind the production of compound substances. Theoretical understanding of this kind cannot, I suggest, be accounted for *solely* as the product of technical, or means-end, rationality.

Developments of the kind described above are only possible where individual scientists have taken it upon themselves to work imaginatively with the data. The first step might be to simply describe the available sensory evidence in a new way, a way which perhaps suggests illuminating analogies with other parts of the natural world. So, for example, Harvey is influenced by the image of the Copernican solar system, with its centralised sun bestowing light (and, in Earth's case, life) on the orbiting planets; the image inspires the idea of viewing the heart as the centre for the body's blood circulation. A further step towards developing a new theory might involve a sudden shift in how we view causal relations: we may, for instance, start to look at something as related to phenomena which were previously thought irrelevant to its occurrence. An imaginative shift of this type opened up the possibility of explaining the ocean tides with reference to the moon's gravitational pull.

Nature, as we know, does not helpfully label things to indicate their causal inter-connectedness but rather it leaves us to try to discern as best we can the hidden relations between things. We 'fill in' the causal connections according to the manner in which we choose to describe and structure our accounts of nature – although we can never know for sure which of our explanatory structures reflects nature as it really is. Thus, there is an open-endedness to all our attempts to study the world around us. There is always the possibility of casting our experiences in a different light, one which reveals or explains more than before, so that we can never decisively rule out all the alternatives without becoming dogmatic. The problem faced by theorists of *all* types is this: there simply is no single solution decisively endorsed by the evidence of the senses[13], the evidence which the object of study (the natural or social world) gives to the enquiring subject. The senses give us a start, certainly, but then other faculties have to go to work to interpret the data and to organise them into a coherent conceptual system. The more complex the phenomena scientists seek to explain, the more imaginative will be their response and the greater will be the diversity of explanations and interpretations of data on offer.

In disputing even the value-neutrality of *natural* science, I am suggesting that we cannot give an account of how we come to know in this area without reference to the whole of the scientist's belief system, including value judgments. This is because the imaginative faculty, which plays a part in piecing together explanatory theoretical structures, is a product of the whole personality, a whole life-time of experiences and reactions. Imagination does not consistently draw its strength from some one part of the self, and even if it does, we quite possibly could not know which part that is. Imagination can have its source out of sight and out of reach of conscious awareness, and indeed this is the very characteristic which renders this potentially productive faculty so mysterious and so mistrusted by those who have sought to set science on firm, more intellectually *visible*, foundations.

The need for creative interpretation in science has led to a number of difficulties, not least for those wishing to maintain the claims of value-free objectivity. Conceding any full-blooded role to the imagination in the processes of theory construction and development has been associated more recently in the philosophy of science with conventionalist and relativist schools of thought.[14] To the extent that subjective factors are acknowledged to influence theory, and our theories in turn influence what we see as facts, then it seems we cannot count on our facts in the way many of us thought we could. They begin to look 'perspective-relative'. Thus, even if the concept of 'efficiency' were not controversial (and it certainly seems less so in the natural sciences[15]) the concept of a 'means' itself is in no way independent of the subjective perspective since the perception of facts (though not, of course, the facts themselves) depends upon what the subject brings to its research.

There are a number of key differences between the natural and social sciences which have led to the conviction that the natural sciences at least can be value-neutral. Firstly, in the case of social reality there is some recognition that what people think of it *can* change the nature of that reality. This generates problems because social science itself affects people's beliefs and responses and therefore becomes another factor shaping the reality it is meant to explain. To some extent, therefore, the social scientist must regard him/herself as part of the thing being studied, as one of a number of potential causal influences. In the natural sciences, on the other hand, it seems quite appropriate to think of the object of study as entirely uninfluenced by the beliefs and values of the enquiring subject. This helps promote the idea that the natural scientist can achieve greater intellectual detachment. Secondly, because the facts of the natural world are what they are independently of what any person thinks, it is easily assumed that in bringing these facts to conscious awareness *their* value-neutrality is passed on to the knower such that we think we can come to know something in a value-free way. We move from the idea that some facts in themselves are value-independent to the quite distinct idea that our knowledge of these facts, and the processes involved in coming to know them, are value-independent also. This is a mistake, rooted in the idea that the acquisition of objective knowledge requires the subject's passivity or transparency. To be objective, on the traditional, Cartesian view, requires the subject to 'get out of the way', to refrain from introducing anything which might be construed as subjective into the investigative process. Thus, the assumption of 'subject transparency' (which we owe to Descartes' influence on science and to his particular conception of objectivity) has allowed us to make the move from

21

value-neutral facts to the belief in the possibility of a value-neutral method for *discovering* the facts. The presence of individual scientists employing such methods is viewed as coincidental; rational, scientific investigation is taken to be as impersonal as the mechanistic world which is to be the object of study.

I mentioned earlier that the presence of subjective factors influencing theory development has had far-reaching effects on our perceptions of science, reflected in the *philosophy* of science by an increased interest in relativist and conventionalist positions. Once theory preference is shown to be 'tainted' by the presence of the subject, then the facts which those theories uphold are 'tainted' also. The result has been a recent loss of confidence in claims to a universally-valid objectivity. What we once called 'the facts' now begin to look merely 'perspective-relative'. What I wish to do in the chapters that follow is to argue that knowledge need be no less objective for being gathered from an individual, value-laden perspective. Some perspective on the world is, after all, unavoidable but this does not mean we should assume that no viewing-point is any better than another. We need the freedom to imaginatively explore alternative perspectives, to get involved with, and to compare them, whereas the value-neutrality requirement specifically rules out the possibility of any far-reaching perspectival re-evaluation. It has been commonly accepted that the attainment of objective knowledge requires no radical re-adjustment on the subject's part, except the adjustment in favour of intellectual detachment. I will be arguing for an alternative way of conceiving of rational enquiry, one which does not require the subject's imaginative passivity, but which instead demands imaginative *involvement* to assist in the interpretative process which is essential to theory development. Unless some alternative way of viewing the process of rational knowledge acquisition is considered, we will find ourselves left with the uncomfortable conclusion that the exercise of our rational faculties (even through the medium of natural science) will simply get us no closer to understanding a mind-independent, objectively real world. If there is indeed such a world, we will then remain alienated from it.

Notes

1. Vienna Circle: the name given to an influential group of logical positivists, the most famous members being Schlick, Carnap and Neurath. Later joined by AJ Ayer in England. Popper, whose ideas will be discussed in Chapter 3, remained on the peripheries of this group.
2. I am referring obliquely here to the later mutation of positivism into conventionalism.
3. Weisskopf, W.A., *Alienation and Economics*, New York, 1973.
4. Ayer for instance (in 'I think therefore I exist', in H. Morick, *Introduction to the Philosophy of Mind: readings from Descartes to Strawson*, Glenview: Scott, Foreman, 1970) shows that we are deceived by the structure of our language (the self-reflexivity of the personal pronoun) into drawing substantial ontological conclusions concerning the existence of a mental substance.
5. Mirowski, P., *Against Mechanism*, Rowman and Littlefield, USA, 1988.
6. Hayek,F.A., 'The Mirage of Social Justice' in *Law, Legislation and Liberty*, Routledge, London, 1976. Hayek makes this link explicit, arguing that *because* the market is an impersonal mechanism (allegedly morally akin to a natural phenomenon) it makes no sense to complain that its economic consequences are unjust, however unfortunate they are for some. The consequences of attempts to interfere with the market, however, *can* be viewed as unjust, since then some specific person or persons can be held responsible for intentionally bringing them about. Hayek trained as an economist and his ideas have had a profound influence on political philosophy, which is still dominated by the search for impersonal mechanisms to determine distribution, and an associated obsession with the concept of *procedural* justice. (Rawls' concern for 'pure procedural justice' is an obvious example.)
7. This is a reference to the Marxist thinkers of the so-called Frankfurt School which includes Horkheimer, Marcuse, Adorno and Habermas. They believe that scientific knowledge is neither politically nor morally neutral because inherent in its epistemological orientation is the desire to manipulate nature (or society). For critical theorists, the proper goal of knowledge acquisition is emancipation.
8. Apel, TW., *Against Epistemology*, Basil Blackwell, Oxford, 1982.
9. Habermas, J., *Knowledge and Human Interests*, Heinemann Educational, London, 1972.
10. Popper, K., *The Open Society and its Enemies*, Routledge and Kegan Paul, London 1966.
11. Keat, R., *The Politics of Social Theory*, Basil Blackwell, Oxford, 1981.
12. Camus' philosophical concerns illustrate this point well. In his essay 'The Absurdity of Human Existence' (in Klemke, *The Meaning of Life*, OUP, 1981) he suggests that the most important question for philosophy is that of suicide. In common with other existentialists, Camus sees the apparent meaninglessness of human existence as the biggest problem to be overcome. However, existentialism itself can be seen as the product of a liberal culture dominated by an impoverished conception of rationality.
13. The Quine-Duhem Thesis ('Under-determination of Theories by Data') supports the claim that no single theoretical solution is *decisively* endorsed by the evidence of the senses. Quine states that 'theoretical statements impinge on reality only at the periphery'.

14. Kuhn, for instance, upholds the idea that there is a subjective input in scientific research. He agrees that there are sociological and psychological aspects to the development of science but has problems reconciling this recognition with the idea that science can progress rationally. He seems at times unwilling to embrace the relativism to which he is surely committed.
15. This is because the scientist's actions are not directed at human material in a social setting but at inert matter often within the confines of a laboratory.

2 Knowledge and the Knowing Subject

1. 'Knowing' as a Form of Interaction

In the last chapter, I argued that traditional conceptions of rationality serve a destructive purpose. They have presented us with a view of ourselves as individuals isolated from the objects of knowledge; from other people, with their private concerns and needs, and from the material objects of the external world. We are then depicted as struggling to bridge this gap, searching for conditions of certainty in knowledge to reassure us that we are in touch with reality. These conditions, if met, would ensure that we have crossed the gap safely between 'in here' and 'out there'. The world in which we think we live will have been confirmed for us as real.

Both rationalist and empiricist traditions in philosophy have had this effect, of setting the knowing individual at a distance from the world. The layperson coming to philosophy for the first time is very soon persuaded, or perhaps embarrassed, into abandoning the naïve realist position. The empiricist tells us that we may admit as data only what is given in experience; the rationalist, on the other hand, sets out to reconstruct reality from only what appears as necessary to the reasoning mind, but on neither of these accounts can the bewildered layperson readily pass beyond the premises of an argument to the reality with which s/he would otherwise be confidently acquainted. If finally persuaded to abandon naïve realism the enquiring subject will have begun the journey down the path towards scepticism, or at any rate, will have experienced a disorientating jolt as the gap opens up at her feet, separating the subject from the world she thought she knew. To close the gap between ourselves as perceivers and the world as object of perception I have thus far tentatively suggested that knowledge be seen as requiring the act of imaginative involvement, an act which unites the perceiver with what is perceived. At the heart of this view is the idea that 'to know' is not the attainment of a thing called knowledge which is to be found in books or inside the head and whose presence (or absence) can be established without doubt, but rather that it is the assertion of a relationship of involvement with the object of one's awareness. I believe that an interactive account of knowledge along these lines is necessary to overcome the feelings of alienation and separation which have been encouraged by non-naïve realist approaches in philosophy.

Non-interactionist approaches to knowledge have, however, been hugely influential and, I would suggest, damaging, particularly in the field of social scientific research. The widespread acceptance of the value-freedom doctrine throughout the natural and social sciences has been one of the most significant outcomes of adopting a non-interactive view of how we come to know. It has been commonly supposed that the choice of values is entirely arbitrary because the 'rational' individual is seen as imaginatively detached from the world in which some things may really be valuable. The state of intellectual detachment which is thought necessary for objectivity simply isolates individuals from the world and from one another and leaves them with no clue as to how to respond appropriately to the situations in which they find themselves.

Acceptance of the value-freedom doctrine is based on the conviction that the world to be studied is an objective (in the Cartesian sense), impersonal mechanism to which values and meanings are arbitrarily attached. We can therefore only hope to understand such a world as it really is by freeing ourselves from subjective evaluations and by viewing with impartiality the 'cold, hard facts'. This, as I have argued, is a view of understanding based on a limiting, ultimately destructive conception of rationality. As already argued in the last chapter, not even an acquaintance with the so-called 'cold, hard facts' is guaranteed on these grounds, when we take into account the need to interpret and organise sensory evidence into complex theoretical structures.

The relationship between our knowledge of fact and our knowledge of values is by no means a straightforward one, though it has been supposed to be so because it is assumed that the *concept* of a 'fact' can be readily separated from the concept of a 'value'. Clearly, facts concern what *is* the case, whilst values are related to attitudes towards facts; they are determined by what we would *like* to be the case. The problem of keeping fact and

26

value separate, however, only arises when we ask the question: how do we *know* what is the case? Can we enquire into the factual nature of the world without our attitudes influencing what we find? Obviously there are many who think that we can.

Empiricists have typically believed that the facts of the natural world can be known to us by following certain impartial procedures: reality comes to us through the medium of sense experience but it must then be reconstructed solely out of our sense-data building blocks if we are to acquire true (pure) knowledge of it. For the rationalist, on the other hand, reality must be pieced together through the use of arguments concerning its necessary structure. On both accounts, however, the starting point is the assumption that a gap between subject and object needs to be bridged by moving in secure steps from what is certain or immediate to whatever lies beyond. In both these traditions, we begin with the separation of subject from object and then handicap the subject in that we are assumed not to have any capacity for direct, interactive understanding of the objects to be known. The most fundamental assumption at work here is that to establish the possibility of knowledge we must first find a solid foundation stone upon which the edifice of human knowledge is to be raised. There is an analogy at work here taken from one of the most long-standing traditions in scientific thought.

Many scientists believe that if they can find the most basic constituents of the universe (be they atoms, quarks, or whatever) they will at long last have found the key to understanding everything. If the properties of the most basic elements can be understood, it is surmised that we will be able to move from there by logically and empirically secure steps back to the full, macroscopic picture of reality as the ordinary man or woman knows it. Our knowledge of the whole of reality is built up, on this view, from a sound knowledge of some one part of it, a part that is thought to be more essential to explaining and understanding the universe than any other. In more recent years, this view of explanation in science has been challenged. An alternative approach has been proposed, labelled the 'bootstrap' method, based on the idea that no part of the universe has a more elementary role in explanation than another: all parts are inter-related and they 'lift themselves into existence by their own bootstraps'[1]. This being the case, reality can be understood only as a whole and understanding will ultimately elude us if we seek to take the 'whole' apart and try to reconstruct it on the basis of some constituent element. I now wish to suggest that philosophers have made a similar assumption to traditional scientists in attempting to found knowledge in general upon a single, supposedly secure, bedrock. Perhaps, in doing so, philosophers have made the task of explaining how we come to know an impossible one.

Once we deny the possibility of a direct involvement with the objective world through our consciousness of it, we cannot then restore the confidence in our knowledge of reality through any other means. Knowledge cannot be seen as made up of discrete items, each of which has met some theoretical standard of indubitability before being admitted as knowledge-proper. Rather we should think of 'knowing' as standing in a certain *relationship* to the world, a relationship which obtains between the perceiver and the perceived, and which can be stronger or weaker, producing more or less effective action. Successful action (for instance, having the results the actor intended[2]) is one indication of the individual's standing in the appropriate relation to the world, but if action is persistently unsuccessful, or even harmful, we will have reason to say that the individual is not acquainted with the environment in which he or she has been acting. One of the reasons for such repeated failure may be that individuals have set a distance between themselves and the world, alienating themselves from it, refusing to 'get involved', and thus denying themselves the knowledge they need to succeed. Ultimately, the only test there is for what counts as knowledge is the practical one.

'Knowing' or 'to know' should be seen as another kind of action, the act of becoming imaginatively involved with one's environment in at least some respect. The persistent failure to exercise one's potential for knowledge is equivalent to pushing some aspects of reality away from conscious awareness and insofar as the subject is diminished by this act (the result of an alienated choice), s/he must be deemed irrational.

Thus, rationality, given full expression, may be understood as consisting in the reality-encompassing act which unites, or directly involves, the knower with what is known. If we return now to Weisskopf, we find that he proposes a very similar view to the one proposed here: a conception of rationality which he calls 'encompassing' or 'ontological' rationality of which the afore-mentioned 'technical' rationality (from Chapter 1) is merely a dependent part. Moreover, Weisskopf suggests, it is a part which has usurped the place of the whole.

Weisskopf explains encompassing rationality as 'the structure of the mind which enables the mind to grasp and transform reality'[3], but it is not clear, initially at least, how this is something more than technical rationality itself which has as its goal the control and transformation of reality. For Weisskopf, rationality is something more than what enables us simply to control or predict. Encompassing rationality enables us to comprehend the world in aspects other than those which relate to our control of it; it is a form of rationality concerned with the attainment of a fuller understanding, and (with regards to people) a fuller sympathy. The exercise of such understanding enables us to interact with the world fully at all levels, that of manipulating and controlling our environment, of empathising with other individuals, of responding constructively and creatively to improve the physical environment, our society, and the lives of others. Thus, when we are uninhibited in the exercise of our rational faculties we exist on all levels – the physical, the social, *and* the moral – and as such lead full and unalienated lives.

By contrast, with only technical, or instrumental, reason to guide action, we can interact with the world in only a very restricted manner, a manner which does not do justice to our rational faculties. Of course, the sceptic will want to deny that (for instance) inter-acting on a moral level is part of what is required by rationality. If the moral sceptic repeatedly fails to behave in a way which shows any concern for others, s/he does not typically accept that s/he is failing to exercise certain innate capacities and is thereby alienated. After all, the sceptic is still exercising the capacity to choose in *refusing* to identify with the interests of others and so long as we have free choice, it is commonly thought, how can we be alienated?

With the refusal to acknowledge a broader conception of rationality, one which provides for the possibility of purposeful ethical debate and the prospect of agreement upon objective truths, modern moral philosophers have been left with an apparently arbitrary, irrational act of choice at the basis of their account of moral thinking. The prescriptivist, following Hare[4], may go a long way towards providing a broad basis for agreement on moral principles by showing that the logic of moral language is such that we must be sincerely prepared to accept the consequences of our universalised prescriptions, regardless of whose position we imagine ourselves to be in. Perhaps if we did adopt this approach to moral thinking, we might all agree to pursue those courses of action which (for example) minimised suffering, but Hare cannot rule out on the basis of his account of moral judgments the claims of individuals who reject these particular prescriptions. It is quite possible for someone to universalise the prescription: 'Let the elderly die' and sincerely wish this to be so even if s/he were among the elderly. If it so happens that this awkward individual gains more satisfaction from the death of the elderly than the sum total of dissatisfaction of those who have to die, then prescriptivists cannot hope to engage in any further debate to change this person's mind.

A second school of thought has sought to replace the arbitrariness of individual preferences in selecting principles of action with the preferences of the collectivity or linguistic community. On this account[5], whichever individual behaves in a way inappropriate to modes of behaviour laid down in the rules and conventions of the community has thereby violated an 'objective' moral code. 'Objective' here is used in the sense which implies that the moral code is determined independently of the individual and his/her subjective preferences. However such principles of behaviour are not objective in the true sense which implies that these principles can be rationally assessed from a perspective outside of the community. A moral objectivist in this stronger sense may wish to reject moral codes of practice of the entire community, seeing that the code of behaviour which prevails is one that has merely arisen from the preferences of society as a whole, and not one which is based on systematic and rational ethical debate.

The point that these two very general schools of thought have in common is that ultimately moral decisions are based on a non-rational choice, whether it be of an individual or of the community as a whole. By 'non-rational' it is meant that the choice made is a matter of taste, of convention, or of habit and custom, and as such not disputable. Thus we may say 'I/we like this' and it is not appropriate to say 'But why?' It is assumed that taste cannot be swayed by any form of rational argument and it is for this reason that the possibility of an *alienated* choice is denied.

It is only when we view knowledge, including moral knowledge, as requiring some form of imaginative interaction that we can begin to understand why some choices – like those of Hare's 'fanatic' – really are repugnant to reason.

2. The Individual and Claims to Knowledge

In *Legitimation of Belief*[6], Gellner criticises conceptions of rationality such as the one proposed by Weisskopf[7] on the grounds that if one argues for the possibility of knowledge by appealing to a supposed natural harmony between the structure of the mind and the structure of reality then there can be no grounds for rejecting any claim to knowledge one might wish to make '...because the mind is part of nature, *any* principle it employs in the interpretation of nature must therefore be sound.' Obviously neither Weisskopf nor myself wish to make such a rash claim but fortunately I do not think Gellner's criticisms need to be taken as final and crushing. What Gellner *does* do is to highlight the need for an account of what, on such a view, distinguishes knowledge from error. In other words, given that our minds are ready-equipped with a structure which enables us to grasp reality, (ie. given that we are endowed with something like Weisskopf's 'encompassing' rationality) how then is it possible that we make errors? How is it possible that there is such a variety of opinions about what is true?

To respond to Gellner: the assertion that our minds are so constituted as to mirror the structure of reality is merely the necessary precondition for the possibility of knowing objective truths. It is the precondition which grants us the potential to know but does not guarantee that, on every matter, we actually *do* know, that we may legitimately be said to possess knowledge. In Chapter 7, I shall focus on the different types of error and ways in which we can go wrong in trying to understand the world. In this chapter, however, I shall concentrate on the condition of *being knowledgeable*, that is, of standing in a relation to the world as of knower to thing known.

If we choose to view knowledge as a relation between subject and object then there will be a number of important implications, both for knowledge as such and for the knowing subject. Firstly, it is only the *individual* who can bring him/herself into the appropriate relationship with reality, and, as has already been indicated, the individual is always, to some extent, free to decline. The establishment of any kind of a relationship, including the relation of knower to thing known, requires an action, and, like other types of action performed by conscious subjects, it will involve a choice. Thus, we can choose to know or not to know without this affecting the claim that we are endowed with rational faculties, that is, with the *potential* to know. Secondly, the onus is on the subject as the active partner in the subject/object relationship[8] to move towards the object, to imaginatively grasp what it is; the object meanwhile sets the limits on what counts as having knowledge of it. Thus the subject/object relationship which corresponds to that of knowing/being known is uniquely determined by the nature of that which is known and in no way determined by the decisions, or preferences, of the perceiving subject. This being so, there is then the possibility of a subject's making an alienated choice, since choices to know or not to know are measured against the benchmark of an independent, determinate reality.

If 'knowing' is viewed as a special kind of relationship between subject and object, then it is clear that only the individual subject engaged in studying the world is in a position of final authority in the assessment of his or her own knowledge claims. The acquisition of knowledge is a *personal* process and the ultimate benefits of acquiring knowledge are personal also; the more each of us knows, the greater will be the cohesion and predictability of our experiences and the less will be the extent of our confusion and separation from reality. Of course, these criteria (less confusion, more cohesion, etc) do not provide the individual with anything like a basis for certainty in knowledge. They do not tell me that the state I am in now is one in which I am *without doubt* maximally in touch with reality, but I do not think this ought to be taken as evidence that this is not what the possession of knowledge consists in. The most important implication of this view, which suggests that certainty is unattainable, is that our knowledge claims should be constantly subjected to critical re-assessment, that the widest variety of opinions and approaches to study should be discussed and evaluated in as thorough a way as possible without placing any arbitrary limits on what types of belief [9] are relevant to the debate. This is a position very close to that of the critical theorists, such as Habermas, and to the fallibilists who argue for the importance of freedom of speech and open discussion on factual as well as moral issues.

The way in which we think about the concept of knowledge has been dominated by the desire to find conditions for certainty in knowledge. When guarantees cannot be found neatly attached to our knowledge claims the result has too often been despair and/or a retreat to some form of scepticism, but this need not be our response. The lack of certainty and the unavoidable open-endedness to all forms of enquiry present us with an on-going challenge, not an intractable, infuriating problem. Humanity's craving for certainty is symptomatic only of immaturity, rather like an insecure child who wants everything spelled out in black and white, with no grey areas or suspended judgments. As children, we do not really have to think for ourselves – we look to our parents for all the answers – but as adults we have a moral obligation to think about and to challenge what others tell us to believe.[10] Too often the child/parent relationship is simply succeeded in adulthood by a willing submission to new authority figures – experts, leaders of various kinds – who seem to offer the certainties we desired in early life.

Thus, instead of thinking about issues directly, we ask: What does the recognised authority in this area have to say? – What do *other* people think? – What are the laws and conventions pertaining to this matter? In this way, the individual's participation in the on-going struggle to understand the world for him/herself is given up in favour of allowing others (only *seemingly*) to do the job instead. In fact, no-one can really do our knowing for us.

In seeking to understand the world, there are inevitable uncertainties but many philosophers, including Mill and Popper, recognise that the necessarily tentative nature of our knowledge claims does not imply that they are not therefore claims to *objective* knowledge. What I claim to know might in fact be the truth, though I could without contradiction accept my fallibility in making the claim. In general then, the supporters of fallibilism will tend to stress the importance of free and critical debate because of their recognition of the fallibility of any claims, no matter who makes them or however secure they may appear. Fallibilism encourages not only the continued activity and restless questioning of the enquiring mind but also fosters a healthier, essentially critical, attitude towards even the most widespread and popular beliefs. The impossibility of gaining complete certainty in knowledge ought to spur individuals into seeking knowledge for themselves instead of accepting it from others on faith. The impulse to merely accept the beliefs of those who lay claim to intellectual (or other) authority must be resisted if subjects are to remain both rational and responsible.

The personal aspect to legitimate knowledge claims implies that the individual's confidence in his or her beliefs cannot ultimately be founded on the word of whoever is considered to be the reputable source of knowledge. This is not to say that the sensible individual should not attach some significance to the ideas of those who may have devoted years of study to a particular field. Expert opinions serve as valuable guidelines to those less experienced or as yet wholly unacquainted with the issues. The role of experts, however, is to guide others in what can only be a uniquely personal search for knowledge; after all, one can only establish the knower/known relationship for oneself and not on another's behalf. The bottom line is that the novice and the expert are alike in that they have nothing but their own experience upon which to judge the truth of their respective beliefs. They may judge for *themselves only* how adequately their concepts and explanations apply, how fully one set of explanations integrates with others, how much of reality they feel personally *involved with*, in the sense that

33

understanding requires the involvement of the subject in the object of interest. Such an understanding is accompanied by a diminishment of confusion (this follows from the elimination of inconsistency and/or the drawing of more aspects of reality under the light of explanation), an easing of the feeling of isolation and, to some extent, therefore, of feelings of uncertainty also. Authoritative sources play a part only to the extent that the individual is able to recognise that their authority is well-grounded; in other words, that what they say seems objectively right or true or fair, in which case it is *this* insight and not the mere recognition of their authority as such which will play the decisive role in the deliberations of a truly rational subject. At the end of a process of deliberation, individuals can only claim knowledge if they have decided for themselves, rather than simply consenting to what others believe, and even then, their claims – like those of the experts – will remain contestable.

To make the personal aspect to legitimate claims to knowledge clearer let us consider an unsubstantiated claim to knowledge: my claim to 'know' that the explanation of the origins of the universe is given by the 'Big Bang' Theory. In fact, I have little idea what the 'Big Bang' was supposed to be, how it happened, or how it serves to explain the existence of matter, but suppose I claim to know that the 'Big Bang' Theory is true simply because some very reputable physicists seem to have been convinced by it and I feel sure that they must have good reasons for their convictions. What is more, I have a number of friends whom I respect and who also claim to 'know' that the universe began with a 'Big Bang'.

Now, you may ask, how can any of us *non*-physicists (my friends and I) claim to know that the universe began with a 'Big Bang'? Apart from some scanty, lay-scientific explanations from the *Guardian* science and technology pages, we really have little conception of how the 'big bang' explains anything, or, for that matter, what alternative explanations there are. Perhaps if I had studied physics more I might have some basis for making a judgment on the merits of the 'Big Bang' and other theories, but for the moment I would really have to say that I do not know much about the origins of the universe, but I do know that certain physicists have very strong opinions about it and doubtless for good (ie. convincing) reasons. Thus, when in the past I, and others, have rashly claimed to believe in the 'Big Bang' explanation, perhaps my claim should really have been understood as a youthful expression of confidence in the findings of scientists, or even as an expression of the idea (which I *may* have been justified in holding) that the 'Big Bang' Theory had entered into the body of accepted scientific truths and that one day, if I bothered to look into it, I too might confirm that it is indeed a very plausible explanation. These disguised assertions, however, are quite distinct from the claim *to know* anything about the origins of the universe, for I cannot claim to know when I simply do not understand the intricacies of the debate. Such bogus knowledge claims are not entirely unlike the knowledge claims made by those who capitulated under the pressure of the Inquisition: the professed religious belief of former heretics was not based on personal experience of a loving Christian God (quite the contrary) and thus counts for nothing except as a testimony to the power and cruelty of the religious authorities of the time. Of course, intellectual, moral and political authorities of the present day do not tend to employ gruesome instruments of torture (they keep quiet about it if they do) but nevertheless they have their methods which, it might be argued, are all the more persuasive in virtue of their subtlety. People are instilled from an early age with a respect for authority – and, these days, particularly with a respect for science – such that we are prepared to accept the pronouncements of those whom we deem authoritative almost without question. The result is that we stop learning and exploring the world *for ourselves*. Respect for authority can cripple the thought processes quite as effectively as fear, producing an involuntary suspension of our critical faculties – like a reflex action – occurring at the very point where the authority's judgment comes into play.

When Popper speaks of the importance of an open society for healthy intellectual debate, when he refers to the dignity of man in the pursuit of knowledge[11], he apparently has this in mind: that society should be such as to foster the critical attitude, to encourage individuals to challenge accepted ideas and seek understanding for themselves, because none of us can claim to know with absolute certainty. When there remains the possibility of being wrong, or of being only partially right, to close the debate as if our answers are final is tantamount to closing off areas of reality which may not yet be properly known. Imagining that we can achieve finality, or completion, in our knowledge of the world may therefore be severely alienating. However, there is a potential conflict between Popper's ideals that he may not have been aware of. He is a strong proponent of the critical attitude, and a supporter of the freedoms required to foster critical debate, but at the same time he does not envisage a society without centres of *political* authority. There is at very least a tension between the demands of a political authority in a given area and the demand that a rational, autonomous individual challenge and establish the authenticity of any claims or recommendations for him/herself. The critical attitude developed to its full extent seems inconsistent with the recognition of any political obligation. Popper agrees that political tyranny is inconsistent with an intellectually healthy society but he *does* believe[12] that the Western-style liberal democracy successfully reconciles the critical attitude required of a rational agent with a submission to political authority. In later chapters (particularly Chapter 8), I will be examining the plausibility of this position.

In so far as Popper recognises the importance, indeed the primacy, of liberty as a requirement in the pursuit of knowledge, he may be considered alongside another renowned philosopher of science, Paul Feyerabend. Personal liberty may be said to be a focal point of Feyerabend's work, but there are, of course, very important points of contrast with Popper's work. For Feyerabend[13], science cannot be undertaken as if it were an activity free from value-judgments or value-implications because he believes that even at the level of the interpretation of data and of the imaginative selection of some new explanatory hypothesis from amongst the many possible hypotheses, there will be an involvement of the *whole* of the personality of the scientist. The scientist as a unique individual will bring to his/her work a set of preferences, goals, aesthetic and moral criteria which will colour and shape even a description of the so-called 'raw' facts. Given the infinite diversity of human nature, Feyerabend sees no reason why a healthy scientific community should not be characterised by a diversity, or proliferation, of theories. The extent to which this actually *is* a feature of scientific communities is an indicator of how much encouragement is given to expressions of individuality and of how much freedom there is to question and to criticise established viewpoints without fear of reproval.

That such reproval, even ridicule, exists is suggested by the fact that in the sphere of ideas, just as in clothes and music, there are trends and trend-setters; there are people who set the band-wagon rolling and those, more numerous, who are happy to jump on it. Whilst welcoming individuality and the consequent proliferation of ideas, I do not wish to condemn the practice of jumping, or at least of *appearing to jump*, onto band-wagons, for there can be no harm in following up a good idea if that is how it is genuinely perceived. But here we have a problem, for how can we tell if we have been persuaded by the strength of a good argument or swayed by the endorsement of a most respected authority? In other words, how can we be sure we are not jumping on a band-wagon simply to be with the rest, to avoid being left on the roadside?

What this difficulty brings to light is the role that critical debate has, not just between people, between intellectual/political/religious factions, but *within* each individual's own thought processes. The implication of fallibilism, that all ideas are left open to criticism and continual re-assessment, applies equally to our own internal struggles to achieve knowledge. The willingness to question our own ideas, their origins or motivational source, is just as necessary as questioning the ideas of others if communal debate is to have any purpose. It must be the sincere goal of all participants to extend their knowledge of the real world, even if this means admitting to error. In a later chapter, I will suggest that the acceptance of the value-neutrality doctrine in science rules out the possibility of examining the inspirational sources of accepted knowledge claims and, as such, it reduces the likelihood of our advancing any further towards objective knowledge.

Unlike Popper, Feyerabend does *not* defend the possibility of knowledge of objective truths but he does speak of progress in knowledge: '...liberal practice is [not just a fact of the history of science...but] an absolute necessity for the growth of knowledge.' His concern, unlike that of the fallibilists, is not to defend the possibility of objective truths, and of our knowledge of them, but to clarify the nature of the scientist's activity, to make explicit the role of subjective preferences and to justify the role which scientists have generally denied them.

Although conceding that there is a subjective component in the development of ideas, the fact that Feyerabend nevertheless refers to 'progress in knowledge' brings to mind certain similarities with the thesis of fallibilism. Proliferation of ideas (a Feyerabendian ideal) and the free expression of individuality are the pre-conditions for rational, purposeful debate. Feyerabend effectively recommends freedom whilst simultaneously denying that there are any objectively worthwhile ends we should be freely pursuing.

Whilst, like Feyerabend, I wish to recognise the inescapable subjective element in scientific enquiry I do not wish to conclude that the knowledge obtained is itself merely subjective, that when we claim to know something we do not express truths about an independent, objective world. I do not believe that the condition of 'being knowledgeable' refers only to a frame of mind, or to a condition of agreement with others in the same community or

37

culture, but rather to a frame of mind, to the world as-it-really-is, *and* to a certain relation between them. Knowledge is obtained when the relation between the mind and the world is a harmonious one, resulting in action which is thereby rational and coherent. Rationality can only be defined by making reference to the agent and its sphere of activity. It has to be a *relational* definition.

It is a confusion, and a common one, to suppose that because we comprehend the world inescapably from our own perspective, we do not therefore know the world at all but only our perceptions of it. I am trying to argue to the contrary, that we do know the real world, though unavoidably under a certain aspect, the aspect made available to us through the subjective perspective.

As a matter of fact, all human beings possess very much the same bodily faculties but our intellectual histories testify to the diversity of opinions and the variety of possible assessments of 'the way things really are'. The reason for this diversity, for the potential uniqueness of every individual viewpoint, is to be found in the infinite variety of human experiences which, in turn, shape our tastes and preferences and serve to mould the very framework within which we interpret the world. Thus it is conceivable that each one of us perceives some part of the world under different, though equally realistic aspects. If the different aspects are consistent with one another then this helps to confirm that they are indeed aspects of the real, independent world and not the result of fantasies or perceptions hindered by the distortions of ideology[14].

The picture we attain of the world is no less truthful for its having been based upon exclusively human perspectives for we can only understand the world as conscious beings, never as unconscious things (inanimate objects) or, more peculiarly still, as nothing at all. In this sense alone all our knowledge is subjective.

3. Self-chosen Values, Imposed Values and Alienation

Encompassing rationality, which constitutes our full rational potential, allows for our interaction with the world on the moral as well as on other practical levels. If we choose to deny the reality of moral responsibility we are choosing to withdraw from the world under certain of its aspects; we are choosing not to interact in a way which yields understanding not only of the physical realm but of other conscious beings also.

Weisskopf argues that when we interact on the practical level, concerning ourselves only with the prediction and control of nature, we are denying a real need to bestow value and meaning, to give purpose to life. In eras not characterised by the reign of technical rationality and so-called 'value-empty' philosophy, Weisskopf observes that the presence of value-laden belief systems provides the psychological and intellectual security which is lacking in the alienated twentieth century. Values upheld rather than undermined by reason play a part in achieving successful repression (Weisskopf's own terminology). Weisskopf argues that the repression of at least some of our desires and potentials is necessary due to the predicament of existential scarcity.

The term 'existential scarcity' expresses the fact that as finite beings restricted to one time, one place, and consequently to one life-plan, we cannot possibly achieve everything we may want to. Some aspect of our diverse potentials will, in the course of a finite life-time, remain unexploited or under-developed and to this extent some degree of repression will always be necessary. The role of a value system in obtaining successful repression is to provide a basis for choice between alternate life-plans. The value system serves to repress certain human desires and to encourage others.

Weisskopf accuses utopian thinkers of imagining falsely that we could have a society where no aspect of human nature was repressed. Given human finitude, he asks, how could this be possible?

I agree that we must concede that humans are indeed limited in time and space (essentially we can choose only one life-plan) but, in opposition to Weisskopf, I see no compelling reason why the life-plan chosen should not be such as to give expression to all aspects of human potentiality. For instance, violin-playing and looking after one's dogs do not call upon two completely different sets of human capacities. Both activities require patience and some degree of self-discipline if one is to practise and walk the dogs daily. Admittedly, neither activity exhausts the range of human capacities, but then neither of them need take up one's whole life-time.

Human wants are not unlimited (as economists imagine) but neither are human capacities, as Weisskopf supposes. We may in fact interact fully, that is, with our whole selves, whilst engaged in a limited set of activities. The essential factor in overcoming alienation is to do what we have chosen to do with the utmost involvement, or, as Erich Fromm has said, 'with love, spontaneity and enthusiasm'.[15]

It is not the frustration of being unable to participate in unlimited activities which causes human alienation, but the frustration of not being able to do what we choose to do to the fullest extent our nature will allow. If it were otherwise we would indeed be irrational creatures.

Weisskopf is mistaken in pessimistically supposing that repression is inevitable and that the role of belief systems is to secure a successful repression. His error stems from a confusion of the experience of regret with the far more psychologically disruptive experience of alienation. Existential scarcity, as Weisskopf has defined it, only entails the rather unfortunate consequence that we must all experience regret, the feeling attendant upon the necessity of our having to choose one path in place of another.

We may be such that even when we sincerely believe we have chosen the better path, we may reflect with regret upon what might have been. This is a regret at not being able to engage in two enjoyable activities simultaneously: for instance, a happy deceit-free marriage is not possible whilst engaged in numerous illicit affairs. Regret is associated with the unattainability of two conflicting goals and arises from our capacity to conceive of other possible worlds. Regret is to be distinguished from alienation in that the latter occurs when *any* activity we choose to engage in proves unsatisfactory, not because we would rather have the alternative, but because even if we had the alternative it would not involve us fully. Alienation describes a condition in which there are continual shortcomings in the manner of our involvement.

It may be that regret, as I understand it, is a symptom of the deeper problem of leading an alienated life and that it occurs when we are failing to be involved in our chosen life-plan as fully as we might. Thus, the woman who desires affairs does so because she is no longer having a fully satisfying relationship with her husband. On the other hand, it could be that even when fully involved to the extent our rational and emotional capacities will allow we may still experience regret. A person could be certain they are satisfied in devoting their life to medicine but might nevertheless wish to do something to help sick animals. Not being able to do both is the consequence of existential scarcity and it is this which causes regret. Such a person is not necessarily also suffering alienation – on the contrary, I would be inclined to say that if a person did not care for the plight of, for instance, sick animals, experiencing some regret at not being able to intervene personally, then that person probably *would* be alienated. That part of the individual which is capable of sympathising with the sufferings of non-humans will have been subdued, perhaps through inactivity or the inadequacy of imagination and the consequent failure to view things from another's perspective. Again, open discussion and critical appraisal of one's own attitudes and beliefs would play an essential role in re-stimulating a fuller awareness.

The notion of existential scarcity, although not necessarily linked with alienation, does serve to emphasise the need for greater flexibility of options in human life. Since our finitude imposes limits on where we may go and what we may do, it is important that if we are to exercise our potential we must at least not impose any extra limitations upon our options. Diversity in activity is as essential to psychological health as diversity of opinions is to healthy discussion. Indeed the former is a pre-condition of the latter since it helps broaden personal experience, fosters the ability to sympathise, and thus aids the understanding. In addition, a more varied and colourful experience often produces a more fertile imagination.

Weisskopf believes alienation to be a sickness peculiarly severe in modern life. This is because he links it to the disintegration of rationality into bare technical rationality and the subsequent destruction of value systems in the era of science. I too believe that alienation is peculiarly severe in modern life but I do not agree with Weisskopf that alienation was not a feature of the pre-scientific societies, where value-attitudes and reason peacefully co-existed and mutually supported each other. I would argue that value attitudes, if they are not self-chosen but imposed (for instance, by the Church) are just as compatible with leading an alienated life as is the absence of an explicit value system.

Modern life seems particularly rich in the occurrence of psychological disorder (a symptom of alienation) because there is so little opportunity for diversity in activity. Consider the organisation of the labour market, the high levels of specialisation and training needed to pursue a single course of activity, and the great pressure to conform. Everyone, it seems, must be capable of being adequately described as either a 'teacher', a 'student', a 'lawyer', a 'park-keeper' or by whatever job title they possess. The failure to acquire and identify with some such job description is then taken as a sign of stubborn resistance to 'proper' work, a childish refusal to conform to one of the socially accepted categories. Unfortunately, it is the sad fact that many people's lives so lack diversity, often because their jobs are too demanding, that the simple label 'teacher' or 'lawyer' actually fits. It is no surprise that amongst the people who can most readily be identified with their jobs (especially when their job draws on only a very limited range of their capacities) we find a high degree of dissatisfaction and psychological imbalance. A society made up largely of such people – even if it is one which generally prides itself on being 'liberal' – will not provide a suitable environment for healthy intellectual debate. The possibility of ideological distortion and of limited personal perspectives is all the greater in such a society and reduces the likelihood of agreement on what may be considered objective truths.

Society, before the advent of technology, capitalism and extreme division of labour, possessed what Weisskopf describes as a belief system underpinned by the faculty of reason. Indeed, he believes that the belief systems prior to capitalism were a true expression of rationality where the latter is understood to mean 'ontological' or 'encompassing' rationality. In so far as this fuller type of rationality was allowed to operate in these societies people did not, according to Weisskopf, experience their alienation. They underwent successful repression.

On this view, which I would like to contest, it follows that, for example, the followers of St. Thomas Aquinas could be counted among those who had achieved successful repression. They possessed an objective value system, based on revelation and yet arrived at by force of reason. Is it therefore non-sensical to speak of a disciple of Aquinas as being alienated? Weisskopf's argument seems to imply that alienation is really the affliction only of modern humanity. In what follows I will argue that his assessment of the root causes of alienation does not go far enough and, indeed, he seems to be influenced by the very conception of rationality he wishes to attack.

At one point Weisskopf defines alienation as the reduction of man to 'a part of what he might be'. Although the deterioration in the richness of our reasoning powers (or what are acknowledged as such) is one way in which we are 'lessened', there is, I believe, another source of trouble which Weisskopf overlooks.

Certainly it is true that people need to employ the full range of their reasoning faculties if they are to feel satisfied, but this is only true because in using their fully extended powers of reasoning it becomes possible for them to make free and responsible choices. Now if for other reasons we are unable to act upon these choices or, more subtly still, we are continually discouraged from making choices for ourselves, then the fully rational person may nevertheless be reduced to a shadow of his/her true self. The reason it has been possible for the technical conception of rationality to oust and outlaw the broader conception (encompassing rationality) is precisely because people's ability to act upon the basis of their rational deliberations has been denied them. Historically, humanity has never been entirely free to exercise its capacity to choose, to discover the world for itself, and, as with many things, inactivity followed by a complete lapse into disuse leads to loss of confidence and the eventual disintegration of the faculty in question. As with other aspects of our intellectual history, the tradition in philosophical thought which restricts the definition of rationality to cover technical rationality alone, is a reflection of the way of thinking in society at large. In societies where there has often been little encouragement to explore moral (and other) questions, to challenge accepted norms and practices, and where, by and large, people have been employed by those with authority to aid in the achievement of the *authority's* ends, there is little call for a philosophy which does not reflect the limited activities which are required of the human rational faculty. A more extended conception, and use, of reason might have a de-stabilising effect on the social order.

To return to the example of the Thomists, I would suspect that there were those among their numbers who felt compelled to accept a system of values imposed upon them from without, by their upbringing within the Catholic Church. How much genuinely open debate went on? How often did people decide for themselves rather than give way, consciously or otherwise, to the religious, intellectual fashions of the time? Those who were even in part influenced by the opinions of the authorities could not be truly said to 'know' what they claimed to know for there is something – the weight of opinion, a concealed command – which has entered into their assessment of the issues and which, rightly, has no place there. Such knowledge claims are really disguised declarations of confidence in somebody or something else and as such imply a corresponding lack of confidence in one's own rational faculties, symptomatic of the disintegration of encompassing rationality, of the individual's loss of autonomy, and of alienation.

Influenced by a science-dominated conception of rationality which all too readily implies moral relativism, Weisskopf fails to bring out the importance of having any particular belief system, or of obtaining one's beliefs in any particular way, but instead only stresses the importance of having *some* values and preferably ones which contribute to successful repression. He does not explicitly say, but he seems to imagine the possibility of alternative value systems, each of which may be capable of producing successfully repressed citizens, free from the experience of alienation. Thus, Weisskopf takes a view not dissimilar to that of Weber[16] who, out of the three general types of authority he identifies, singles out the 'traditional' type as the most stable. Under this form of authority-organisation, habitually accepted values and norms serve to reinforce the existing hierarchy and provide a basis for people's sense of substantive justice. In Weisskopf's terms, these traditionally established norms satisfy our need as beings in possession of encompassing rationality to interact on the normative plane.

Under the alternative 'rational-legal' form of authority (and it is relevant to add that this is the organisation-type displayed in Western societies) there is only a formal, legalistic expression of justice. There is a tendency too towards imposing roles upon people, not just in the sense of holding a certain office, but in the sense that in one's official function one has a public face and, out of the office, a private face. This personality division causes tensions which increase the discontent and social instability and may result in the disintegration of the rational-legal type of organisation into the third, and most unstable of the three types defined by Weber, charismatic authoritarianism.

The interesting thing, from my point of view, is that Weber does not think any of the above types of social organisation are perfectly stable. He only goes so far as to say which offers the highest degree of stability, but he thinks that both rational-legal and traditional types admit of the possibility of deteriorating into some form of totalitarianism if tensions mount and if the appropriate charismatic leadership appears. This possibility suggests that the mere 'tagging on' of a set of norms or values, imposed by an authority in accordance with tradition, is insufficient to create a society of stable individuals. The reason why none of Weber's modes of social organisation can offer long-term stability is because they are all based on authoritarian principles. Thus, even in traditional societies, the presence of an authority-imposed value system does not satisfy the needs and capacities of rational individuals to determine, and to live in accordance with, their own values.

The idea that dependence on authority, and the willingness to conform to its requirements, is an outcome of an alienated nature is to be found in the work of Erich Fromm, in particular in *The Fear of Freedom*.[17] There, he expounds the theory that the drive to conform springs from a deep-rooted fear of isolation, from the disturbing awareness of the distance between ourselves as a subject/perceiver and the external world, the object of our experience. To reduce this distance we surrender our individuality by taking on the norms of society. Alternatively, if we have more authoritarian leanings, we may ourselves be active in imposing the social norms. Nevertheless this too represents an attempt to overcome the distance between ourselves and the object of our domination, since the latter's will is forced by law or by other means to become identical in effect to the will of the oppressor.

Fromm suggests a solution to the problem of isolation and argues that we reject both domination and conformity as false paths to contentment. They are solutions which provide nothing but a temporary and unsatisfactory means of escape from the pain of isolation. The answer to the problem of isolation, of the separation of the subject from the object, lies in the opposite direction, that is, not in the surrender of our individuality, but in the fullest expression of it.

In behaving as the autonomous, rational beings we are, in acting spontaneously on the world in accordance with our capacity to think and choose, we unite ourselves with the objects of our concern. Understanding itself requires an active involvement with the objects of knowledge and although this involvement overcomes the separation of the knower from what is known, it does not diminish our individuality because we perceive, and know, from our own unique perspectives. Individuality is preserved because 'the self is as strong as it is active'.[18]

Fromm emphasises the importance of spontaneity in action. I take this to mean that our responses, if they are not to be alienated, are to arise from the full use of our rational faculties, taking no account of socially-defined norms, of the appropriate 'role' to be played out, of the opinions of experts or the dictates of authority. None of these factors are relevant as such to the individual's struggle to distinguish truth from falsity and right from wrong.

To restore and re-legitimise all aspects of human rationality, people have to think for themselves and to choose their own goals and values on a critical basis. This does not mean that anything which is sincerely chosen after critical consideration is correct simply by virtue of the procedure employed. It is merely the case that, if this procedure is employed, it is more likely to produce broader agreement because the attitudes of the participants increase the likelihood of their correctly perceiving and comprehending reality. The stress on individualism ought not to suggest that this is a recipe for diversity in ethics. On the contrary, because knowledge consists in the individual's identification with the object of his or her attention, greater knowledge obtained in this way implies enhanced feelings of sympathy towards others and a sharpened awareness of other people's needs and interests. If all individuals seek to employ their rational capacities to the full, a more likely result, on this account, would be stronger feelings of solidarity and a shared, fundamentally outward-looking morality.

Notes

1. Jones, RS, *Physics as Metaphor*, University of Minnesota Press, Minneapolis, 1982, p.116.
2. If rationality can be applied not only to the means for attaining a goal but to the choice of goals itself then it becomes possible to speak of the successful attainment of a goal as nevertheless symptomatic of irrationality and thus not really an example of 'successful' action at all. For instance, I may 'successfully' beat my neighbour to a pulp whilst, in some more profound sense of the word, I might be seen by others as a complete failure as a human being for doing so.
3. Weisskopf, WA, op cit, Chapter 2, p38.
4. Hare, RM, *Freedom and Reason*, Oxford University Press, 1963.
5. An instance of the view I have in mind here, where objectivity is derived from the community is S. Lovibond's *Realism and Imagination in Ethics*.
6. Gellner, E, *Legitimation of Belief*, Chapter 5, p93.
7. He also criticises Durkheim and CS Peirce for possessing what he sees as similarly flawed conceptions of rationality.
8. I am thinking here specifically of the case where what is known is an inanimate object of some kind, and not another rational subject who can, of course, make him/herself better understood to others.
9. By which I mean value judgments as well as judgments on matters of fact. In social scientific discussions, for instance, value judgments are not debated because they are not acknowledged as relevant.
10. We have a moral responsibility to think for ourselves and to challenge what others expect us to accept only on the basis of their claimed authority. We cannot know whether a claimed authority (eg. on moral matters) is justified unless we consider the (moral) question directly for ourselves – in which case we become our *own* final authority.
11. Popper, K, *The Open Society and Its Enemies*.
12. Popper, K, *ibid*, vol.2.
13. Feyerabend, PK, *Against Method*, Verso, London, 1988.
    Science in a Free Society, NLB, 1978.
14. 'Ideology' here is clearly used in its negative sense, denoting a belief system which is flawed, not through unavoidable ignorance, but through choices which alienate individuals in societies gripped by ideology.

15. Fromm, E, *The Fear of Freedom*, Ark Paperbacks, UK, 1984.
16. Weber discusses the three types of authority in *The Theory of Social and Economic Organisation*, Wm. Hodge & Co Ltd, London, 1947.
17. Fromm E, *op cit*
18. Fromm E, *op cit.*

3 Common-sense Realism and Rational Action

1. Popper: his Commitments to Realism and Fallibilism

In the first chapter, I spoke of the pervasive influence of Descartes on the development of the natural and social sciences. I also referred to the positivist tradition with the intention of drawing rationalist and empiricist approaches together as subject to the same fundamental criticism: both have tended to undermine the ordinary person's commitment to naïve realism[1] and so have aggravated the problem of alienation in everyday life. Like Popper, I would like to argue that naïve realism is the only intelligible starting point from which rational enquiry can proceed.

In 'Arguments for Common-sense Realism and Against the Common-sense Theory of Knowledge'[2] Popper observes that all forms of rational thought must build upon a common-sense foundation, not because he feels this is an especially secure (ie. indubitable) foundation, but because this is the only foundation we have if we wish to suppose that knowledge is even possible. We may not acknowledge this, we may even set out persuasive arguments in favour of, for instance, idealism, but Popper's point is that we cannot easily deny that the possibility of our arguing, and the use of language which this involves, depends upon our tacit acceptance of the realist presuppositions of language-use. From the four functions of language-use which Popper identifies (these are the 'expressive', 'signalling', 'descriptive' and 'argumentative' functions), he singles out for emphasis the descriptive function which, he says, entails the idea that language is used to describe a knowable and objectively real world. As human rationality is intimately linked to language-use (with its realist presuppositions) Popper argues that our reasons for accepting a realist view are therefore 'at least as strong as rationality itself'.[3] To deny realism is to deny something intrinsic to the nature of, and essential to the evolutionary purpose of, rational thought. Rationality is only to be understood as the means developed in the evolutionary process by which animals have been able to adapt and survive in a real, independently-existing and often hostile world.

I am sympathetic to Popper's evolutionary approach, linking rationality to the disposition to act appropriately in the objective world, because I cannot see what sense can be made of the notion of rationality without reference to action or, more specifically, to successful action. But in addition to this, I would add that since all action involves our having to make a choice (we even have to make the fundamental choice of survival over non-survival, the continuation of the ability to choose over the ending of the necessity of choice) we have always to guide our actions by reference to goals which, in turn, are selected on the basis of belief systems and values. If rationality's purpose is to foster successful action, rational thought must surely be capable of choosing appropriate values and of determining ultimate goals.

It is never open to human beings to act without consequence, according to no set of beliefs, for without some beliefs we could not be said to be 'acting' at all. We would merely be involved in 'occurrences' rather than what are generally called 'actions' and although physical determinism may still be shown to be true, we can never practically dispense with the belief that our choices really matter. We can never have the subjective experience of being involved in an 'occurrence' as if we are merely stones tumbling down a mountain-side; our physical activity must always appear to us as the result of a necessarily value-orientated decision. If rationality itself has no part in the selection of values then it has no part in distinguishing good decisions from bad, or potentially successful actions from the potentially unsuccessful ones. The notions of success and efficiency (as argued in Chapter 1) cannot be defined *without* reference to values, from which it follows that no successfully accomplished goal, or means thereto, can ever be called 'rational' without qualification. On the limiting conception, rationality would appear to be of little help in the evolutionary struggle.

Where I diverge from Popper on the nature of rationality is over his belief that rationality as exhibited in human beings is dependent upon the higher functions of language, which Popper identifies as the descriptive and the argumentative functions. For Popper, the evolution of rationality in humans is intimately linked to the greater sophistication of human over other animal languages. Thus one might deny that there is anything binding in Popper's realist presuppositions of language-use (for instance, by suggesting that we merely dream we are using language descriptively and argumentatively) and, with that, opt to embrace scepticism. The problem for Popper is that membership of the rational, discursive community can always be turned down if one is determined to be sceptical.

Whilst it remains true that more sophisticated languages have enhanced human rationality, it is a mistake to identify the possession of rationality with the possession of a language with the two higher functions Popper describes. As conscious beings endowed with reality-encompassing rationality and faced with the necessity of making choices, we are already directly involved with the real world. We will not cease to be rational beings simply by denying it, any more than this table will cease to be a table as a result of my refusal to acknowledge it as such. The knowledge that is obtained through our direct involvement with the world is simply communicated but is not *created* by language-use, so although sceptics can refuse to enter the arena of discourse (ie. they may reject Popper's presuppositions) they cannot refuse the acquaintance with reality which they have by virtue of being conscious, interacting beings. It is *these* features which make us rational[4] whether we choose to enter the arena of discourse or not. It is as if the sceptic, in asking why he or she should be a rational naïve realist, demands to pass out of existence in order to choose to re-enter it again on his or her own terms. Since we can never be in a position to make this 'choice from nowhere' it follows that if we reject participation in rational discourse, and also the option of rational activity, we may be called 'irrational' and not simply 'non-rational', as if the criteria of rationality need not apply to us.

The sceptic cannot leave the court-room where he or she will be subject to the judgment of being either 'rational' or 'not-rational'; there is no 'non-rational' verdict because the sceptic cannot choose to escape responsibility for certain types of action or choice. The rationality of *any* action is always assessable in relation to its consequences for oneself, other people, or things in the real world, and even a withdrawal from reality or a denial of the existence of our responsibilities will not distance us from the consequences for which we can still be judged. If the world is objectively real, and really of value in its relation to us and to others, then there is a rational and an irrational way to behave towards it. What is of value can be abused by the irrational person who fails to grasp the nature of reality and the relationship in which it stands to the individual. It can never be rational – or simply non-rational – for an individual to choose not to know more fully the world in which he or she must act and survive.

Popper observes, with many others, that an essential feature of the evolutionary process is to learn from and to adjust behaviour in the light of past errors, and he concerns himself with highlighting the role of language-use in facilitating behaviour-adaptation in human beings. Although he considers himself a naïve realist, Popper does not hold the untenable position that on all occasions we are directly acquainted with, and have infallible knowledge of, what is objectively real. (His commitment to fallibilism has been raised in Chapter 2.) On the contrary, the argumentative function of language has developed precisely because of the possibility of disputing – and subsequently determining – correct descriptions of reality. This implies that at least some of us, some of the time, are *not* directly acquainted with the objective world but that the distortions which affect our potential for apprehension are at least potentially eliminable by means of rational investigation. Common-sense, naïve realism, Popper is willing to acknowledge, does not provide us with an indubitable basis upon which to found our enquiries but, he argues, it is the only plausible starting point we can adopt.

I am again very much in agreement with Popper in so far as he rejects as unnecessary the strivings of Descartes to find firm foundations for knowledge. Such strivings are based on an alienating misconception that there is a gap to be bridged between the philosopher's enquiring mind and the object of the enquiry, and that before this gap can be spanned the foundations of the 'bridge' must be set into solid rock and not into yielding quicksand which would cause the bridge to shift and ultimately fall short of its goal. The image of 'firm foundations' is a persuasive one, and it has certainly been influential in epistemology, but I would like to suggest that it is a false analogy for the way in which reliable knowledge of the world is obtained.

Popper criticises the Cartesian approach[5] and what he calls the 'quest for certainty' and suggests that this quest should now be abandoned. This is consistent with his fallibilist outlook, but whilst it is important to recognise that we may be mistaken in the knowledge claims we make, it is, I believe, equally important to have some basis upon which to distinguish some knowledge claims from others as being certain for practical purposes. Such claims, including the claim to know that there is a real, independently-existing world, containing real objects and real people (even if we are mistaken about the finer points) are claims to knowledge which are 'certain', though not in the sense Descartes means. We may contrast this type of certainty, or security, in knowledge with Descartes' by providing a different kind of analogy. According to Descartes, security is bestowed upon our knowledge claims when we have founded them upon the indubitable; his (implicit) analogy is that of building blocks set upon rock, but we, as living, active beings may prefer a less static, less material interpretation of what gives rise to certainty. In buildings it is undoubtedly the firm foundations that count and which give confidence to the building's occupants, but in life more generally confidence is found to develop as a result of continued demonstrations of reliability and regularity. Thus the analogy of a long-standing friendship might be more appropriate as the prototype for 'that which cannot be doubted'.

A good friendship, unlike good buildings, may begin on unpromising foundations and yet once we have experienced it we cannot say that the feeling of confidence it inspires is any less unshakeable than that which we have in a firmly-founded building. Such confidence arises not from auspicious beginnings, from the security of the starting point, but rather it grows out of the process of interaction with other people, or with the objects in the world, a process which builds up our conviction to the point where at least some of our beliefs are not doubted. Certainty in our convictions relates to the unintelligibility of abandoning a given belief as the basis for future action. Thus, the belief in the reality of objects can never be abandoned without giving up on the very *possibility* of rational action. This is why I say that some of our knowledge claims are 'certain for practical purposes'.

2. Facts, Values and Relations

When knowledge is viewed as that which enables us to act appropriately, it becomes clear that it is inseparable from the notion of having a purpose or goal. In the evolutionary struggle, the capacity to 'know' is what has enabled animals to fulfil diverse goals, primarily that of survival but also other goals related to the quality of life. The fact that knowledge is so intimately linked to goal-orientated activity has, however, been overlooked, particularly within the empiricist tradition which counts as knowledge only what is given in sense experience. This empiricist requirement serves to filter out values which are necessary in guiding action since they are not the sort of thing which is accessible to our sensory organs. The requirement that knowledge claims be founded on sense data is the basis of the empiricist doctrine of logical positivism, but the logical positivists carried empiricism further with their assertion that no statement has even a *meaning* unless it can be shown to refer to what is observable or otherwise accessible through sensory experience. From this it follows that the primary object of scientific interest (for the positivist) will be whatever is material, for only material things which possess extension, shape and weight will succeed in impressing themselves upon the five sensory organs of the human being. To make scientific explanation possible as we know it the logical positivists extended the range of what is meaningful to whatever ultimately refers to what is observable, whether directly or indirectly. This allowed them to give credence to the sophisticated, highly abstract theories of modern physics with which they did not feel practically able to dispense.

It is in this manner, by demanding that all statements at least ultimately refer to what is perceived through the senses, that the positivist defined the conditions for obtaining certain knowledge. It is essentially a 'building block' approach to knowledge, not an interactionist one, and as such I believe that it cannot account for how we actually obtain knowledge. The positivist's sense-data building blocks will not provide the basis for certain knowledge because to obtain these blocks one must first dismantle what is already whole and intelligible. On their own criteria for certainty, positivists cannot reassemble the sense-data in such a way as to connect *securely* with the objectively real world because any such connection would be a *relation* between two things (what I have called the 'relation of knowing' which directly unites the perceiver with what is perceived) and as such its presence cannot be established by empirical means.

For knowledge to be knowledge *of* something the term must be relational, but this does not imply a relativist position because the relation of knowing is not a purely arbitrary one, that is, it is not established by mere convention or by consensus. The relation is determined by the real world and since there is only one real world there is only one possible relation with it which is that of knowing it as it is under any given aspect. It is the relation of knowing which elicits rational and purposeful action.

I take it to be a feature built into the positivist approach that knowledge of something real and external to oneself is ultimately unattainable, as is knowledge of all relations between objects including cause/effect relations,

relations of inter-dependency (giving rise to values, for we rightly value what we depend upon) and the relation of knowing itself. Thus, positivists who claim to be realists[6] and not merely instrumentalists, and also those who retain a place for inductive reasoning within positivism[7], are betraying the main tenets of their philosophy in order to lend to it an aura of common-sense which it does not deserve. It is as if, too late, they realise that their account of human knowledge is not a 'working' account: it leaves them in an alien world with no clue as to how to behave in relation to the material objects they perceive. Nothing in this material world can be of value, except what we arbitrarily choose to value, and the objective world as we are said to 'know' it offers up no suggestions for appropriate interaction. We possess knowledge, as it were, but we can do with it what we like. Knowledge, and our very complex rational faculties, play no part in guiding action; indeed they give no impetus towards activity of any kind and it is only when some action (that is, some *goal*) is arbitrarily selected that our rationality serves a purpose in determining the best means to the chosen end. But I have already argued (in Chapter 1) that this is a non-sensical conception of rationality which feeds upon the broader 'reality-encompassing' conception it seeks to usurp.

For knowledge to have a purpose, for it to be conducive to rational activity, it has to be possible to mentally grasp important relations, such as the cause/effect relationship whose perceived reality suggests that our action will be effective, thus providing the impetus for activity. The apprehension of real though not directly observable relations of dependency between individuals and between individuals (or life-forms) and material objects (the environment) suggests to us possible goals for activity since all activity must be orientated towards affecting these unobservable relations in whatever way is perceived as beneficial. Thus even the positivist, whose emphasis is on the material, sense-perceivable reality, is concerned with affecting humanity's relationship of power over the physical environment, employing the knowledge yielded in scientific research to re-direct the forces of nature. The possibility of control gives to the scientists a purpose for their activities, but what have they grasped by means of their five sensory organs to suggest this or any other purpose? Surely the goal of domination (or, for the positive economist, that of accumulation of goods/economic growth) is no more rational an aim than that of lemmings who opt for jumping over a cliff? Sense-data alone can provoke no response and if this is to be taken as the raw stuff of knowledge then knowledge itself can suggest no activity as particularly appropriate – it will not even suggest survival as a more appropriate goal for lemmings. Some further aspect of reality has to be grasped before we can take part in the world as rational, active beings. The question arises: why have positivist thinkers in the social and natural sciences adopted the goals of material acquisition and the control of nature? I believe the answer to this is a very complex one, touching on psychological, social and political issues. But to grasp the answer more fully, one needs to look back to some of the most influential philosophers of the last century. I have already argued that a non-interactionist account of knowledge, such as that of the positivists, sets the individual at a distance from the world of which s/he is already an integral part. The perceived distance, and the resulting diminution in our active involvement with reality, has led to widespread disorientation and a feeling of purposelessness such as has been reflected in pessimistic philosophies like that of Schopenhauer.

In *The World as Will and Representation* [8] we find recognition of the fact that on a non-interactionist, idealist account of knowledge no motivation for action can be found unless one is perpetually deceiving oneself. When we realise the deception, we realise the futility of the struggle in which we have been involved. The natural conclusion for Schopenhauer is to seek to subdue one's rational, activity-orientated faculties in favour of sublime, passive contemplation and thus to finally achieve annihilation of the ever-active will. Schopenhauerian philosophy represents a denial of life so outrageous that Nietzsche's ardent life-affirmation appears as an understandable reaction against it. Nietzsche however does not counter Schopenhauerian pessimism by providing us with a less alienating, realist account of knowledge. On the contrary, he rejects the possibility of our statements corresponding directly with an objectively real world and argues that there can be nothing but our interpretations of the world which are 'true' in no deeper sense than a pragmatic one: 'Truth is that sort of error without which a particular class of living creature could not live.' [9] Given his anti-realist position Nietzsche is consistent in recognising that there could be no order or purpose in the external world. The world in itself appears to him as valueless but, he argues, this is no cause for pessimism since it is also the case that a valueless and indifferent world is not hostile to our aspirations, that is, to whatever order we may wish to impose upon it. Thus Nietzsche introduces the key concept of 'Will-to-Power', an impulse to impose upon essentially chaotic, meaningless reality a form and structure which makes it congenial to our ends.

Nietzsche's response to what I see as basically a problem of alienation is the one shared by many people in technically-orientated capitalist societies, where even religion has declined and offers no reassurance or guidance to lives without purpose. The assertion of man's 'Will-to-Power' appears as one possible response to the perceived separation of the knower from the world he or she seeks to know and to the feeling of isolation which this separation provokes.

William Leiss observes in *The Domination of Nature*[10] that the apparent indifference with which the laws of nature, and the laws of economic behaviour, operate casts the relationship of humanity with its environment unavoidably in the context of domination. Humanity must either submit to the blind operation of natural laws or else must attempt to master them for itself. Whilst I agree that this is the response we have chosen, it need not have been our choice since it is always open to respond to the vision of an indifferent and purposeless world by withdrawing from it entirely, as Schopenhauer recommends. However, I believe that for psychological reasons, humanity in general has chosen domination and self-assertion over withdrawal because we cannot so readily subdue our active nature. (How many of us have the potential to be ascetic hermits who, in their sublime contemplation of nature, forget to eat and subsequently die?[11]) At least in asserting control over the environment we may be commended for affirming life, for not denying our nature as creative, active beings, and for not giving ourselves up entirely to self-destructive pessimism.

A psychological explanation of the human drive towards domination has been provided by Erich Fromm, who interprets it as an effort to overcome feelings of isolation (see Chapter 2). Specifically, Fromm speaks of man's domination of man but I think what he says serves as well to explain man's impulse to dominate nature since we perceive not only a separation between ourselves and others, but between ourselves and the external world generally. Attempts to compensate for feelings of isolation have taken the form, in politics, of exercising authority over others; of subjecting oneself to political authority;[12] of mastering the laws of nature in order to manipulate it; and, within the social and economic sphere, of creating and acquiring unlimited amounts of material wealth to enhance feelings of possession. Possession, like domination, draws external objects closer to us and reduces the feeling of being separated, or of being alienated, from them. Thus, in capitalist society, property rights are paramount, the goal of economic growth goes largely unchallenged, and the aim of scientific research as being the control of nature (leading to the control of man in political struggles exploiting such power) is all but taken for granted by scientists.

I began this section with a brief discussion of what I felt to be an inconsistency in the positivists' position, an inconsistency which I then tried to explain (a) by considering what would be the unpalatable, though perhaps more consistent, Schopenhauerian alternative, and (b) by considering the psychological motivation for the positivists' suggested orientation for human activity. On the evidence of the senses alone – the only evidence admissible to the strict empiricist – no action can be either justified or discredited. In short, no practical guidance is given and yet the positivist scientist sees fit to select the goal of environmental control and the positivist economist that of economic growth. The selection of these goals, or of any others, implies their unacknowledged acceptance of the reality of at least some unobservable, though practically demonstrable, relations of power and interdependency in the objective world. For action to be perceived as purposeful it has also to be accompanied by the conviction that it will, or might be, effective and, furthermore, that the effect will not simply be an alteration in the way the world is, or appears to be, but will be a more significant alteration in the *relationship* between the world and the person acting.

To clarify, it might be helpful to consider an example. Suppose we ask our neighbour why he is erecting a ten foot high fence down the centre of his garden and he replies (not facetiously) that he just felt like having a ten foot wooden fence impressed upon his senses every morning when he looks out of the window. It is not that he finds it attractive or even useful for anything; on the contrary it provokes no positive feelings in him at all. We might be forgiven for thinking that our neighbour is not entirely sane. On the other hand, if this man were to tell us that he was building a fence in order to serve as a wind-break to protect the vegetables he was going to plant, we would begin to see that his action had a purpose. We can now begin to understand his action *as* action, that is, as essentially goal-orientated and therefore to be distinguished from natural occurrences. In the case of the insane man's explanation we are more inclined to equate what has happened with an occurrence in the physical world. We might say, for instance, that the insane man has lost his capacity for choice and has become subject to the whimsical distortions of his diseased brain. If his condition is generally recognised as one of insanity he will be exempt from moral responsibility, just as is the stone which, disturbed by earth settlement, rolls down the mountain-side and injures a man.

The crucial distinction which this example brings to the fore is that between purposeful activity and purposeless occurrences. It also shows that what makes an action purposeful is something beyond what can be discovered in the examination of sense data. When the fence is viewed only as a material object which impresses itself upon my neighbour's senses it cannot be said to be the result of purposeful activity because the object in itself, its height, weight and texture, cannot inspire purposive action. A purpose for the construction of such a fence only arises when we consider such things as the object's impact on the environment, in so far as it impedes the flow of air; the man's need for food, coupled with human dependency on the environment to provide this food; the presence of adverse factors in nature which make it necessary to adapt either the environment or ourselves. Thus it is the particular interaction of any object with the environment and, ultimately, with acting, goal-orientated life-forms, which bestows upon the object a value and may lead to its incorporation in a wide range of purposive activities.

Life-forms, in contrast with dead matter, have the capacity to adapt, for which end we are equipped to a greater or lesser extent with faculties which enable us to grasp, through inference, strictly unobservable relations of inter-dependency. Adaptation is the process by which animals adjust the balance of power between themselves and the environment, tipping the balance in their favour either by developing their immunity to adverse circumstances or else by controlling nature and averting possible disasters. Humanity has tended to favour this latter method but has, in the course of the evolutionary struggle, almost lost sight of the original purpose for tampering with the balance of power; the purpose was not to gain absolute power for ourselves, as if nature were a dangerous enemy who had to be forced into submission. As Nietzsche observes, nature is neither our enemy nor our friend but is, on occasion, as inadvertently kind to us as on other occasions it can be inadvertently cruel. What the human animal who wishes to survive has to recognise is where the correct balance lies in our responses to the environment. A recognition of what is of value, of what stands in a relationship of dependency with living things, is a necessary element in the accurate comprehension of the world as it is. Without the possibility of comprehending what is of value, we could not act rationally and may even unconsciously work against the attainment of some of our most important goals. The example of humanity's destruction of the ozone layer upon which many life-forms depend for protection against the sun's more harmful rays illustrates only too well how our lack of willingness to comprehend reality fully leads us to what must surely be classed as irrational action.

Under the influence of positivism and its limiting conception of rationality, scientists have adopted the goal of extending humanity's power over the environment. Because they have failed to acknowledge the reality and nature of the complex relations of dependency that exist between life-forms and the environment, because they have focused on the sensible, *material* properties of the world to the detriment of the unobservable (but inferable), they have tended to overlook those aspects of reality which, had they been properly apprehended, might have caused some revision of their choice of goals. The aims of controlling nature and, in the social sphere, of material acquisition, are based on (badly informed) self-interest rather than a fuller perception of our place in, and relationship to, the world we inhabit.

Rational beings, faced with the necessity of acting in one way or another, must develop some picture of the how things are related if their actions are to have any purpose. However, each acting individual can only infer these relations for him/herself in a way determined by the totality of the individual's belief system. As we know, nature does not obligingly yield up its own self-descriptions or present events to us neatly labelled as 'cause' to some 'effect'; we have to figure all this out for ourselves and we do so by drawing new data into our existing theoretical structures. We do not directly observe, but we can infer the presence of a causal relationship (as well as other types of relationship) with reference to an existing world-picture, one which will implicitly suggest, or rule out, certain connections. Thus, the causal inter-connectedness of the world is not something which can ever be unambiguously and publicly demonstrated in a way satisfying to the positivist, since the ability to apprehend relations resides firmly with individual members of the species. In this sense alone the apprehension of relations is subjective; it involves a personal effort at interpretation, but this need not imply that there is no correct interpretation, that what can only be *subjectively* apprehended cannot also, on occasion, be objectively real. The subjectivity of the process of assimilation should in no way imply that there is no possibility of an individual going wrong. We may still be said to have wrongly perceived the nature of unobservable relations because the relations themselves are real and are uniquely determined by real physical objects and by the interactions of various dependent life-forms.

Our understanding of nature has been distorted by scientists' refusal to recognise that understanding from the perspective of living, acting beings, who are themselves exerting influence and giving new aspects to nature, is not illegitimate. If we are to avoid the disasters which will result from our continued effort to dominate the environment, to turn it into our personal power-house to provide the means of dominating even our own kind, we must begin to see the world and its dependent population in a new light, not as things which are alien to us and to be feared and subdued, but rather as a reality containing things of value and of which we are an integral, value-creating part. We need not alienate ourselves from this world and then choose to respond to it aggressively. Instead we may choose to involve ourselves with it more fully, adopting a creative role in place of the negative, destructive one.

Notes

1. There are, of course, some prominent examples of naïve realists within the empiricist tradition. For example, Thomas Read, CS Peirce.
2. Popper, K, 'Arguments for Common-sense Realism and Against the Common-sense Theory of Knowledge', from the collection of papers in *Objective Knowledge: an evolutionary approach*, Clarendon Press, Oxford, 1972. The argument that putting the case for idealism presupposes realist presuppositions of language-use is found on pp40-41.
3. *ibid* p41
4. 'Rational' is used here to classify a group with the capacity or potential to reason, as opposed to the word's use in identifying specific behaviour which is *actually* rational (that is, where the rational potential is being realised.)
5. As exemplified in Descartes' *Discourse on the Method*, in E. Anscombe and P.T. Geach, eds., *Descartes' Philosophical Writings*, Thom. Nelson & Sons, 1954.
6. For example, Schlick, who argues that positivism is consistent with empirical realism in the paper 'Positivism and Realism', found in *The Philosophy of Science,* eds., Boyd, Gasper and Trout.
7. For example, Duhem.
8. Schopenhauer, A., *The World as Will and Representation*, Dover Publications, New York, 1969
9. Nietzsche, F., *Unpublished Notes*, p814. Whilst I suggest that the only test for the truth of a statement is the practical one, my position differs significantly from that of Nietzsche. For Nietzsche, truth is defined relative to man's chosen goals and there is no knowable, independent reality against which to measure the wisdom of such choices. In contrast, I suggest that the degree of certainty we attach to our knowledge claims depends upon their efficacy, but the *truth* of such claims is not the *result* of their efficacy but of their corresponding to an independently existing state of affairs.

10. Leiss, W., *The Domination of Nature*, George Braziller, New York, 1972.
11. Schopenhauer, op. cit, vol.1, 69, pp400-402.
12. Fromm talks only of the willing submission to tyrannical authority. However, I feel that the willing submission to any authority other than one's own has, at very least, to be explained.

4 Alienation and the Social Scientist

1. The Authoritarian Assumption

In this chapter, I wish to suggest that in accepting the so-called 'objective', value-neutral approach of the natural scientist, the person who seeks to study social, and specifically human, phenomena has made the task an impossible one. Whilst natural scientists have escaped the worst ravages of ideology[1] and manage to attain some measure of agreement with their colleagues and with the scientific community generally, social scientists find their discipline wracked by dissent over even the most fundamental questions and, significantly, over what one would expect to be easily established matters of fact.

Perhaps the most problematic area for the social scientist is the analysis of causal relations that exist between individuals, institutions, or economies, because the analysis of these unobservable (though inferable) relations depends upon the perceiver's own framework of belief. Relations, such as those of causal connection or mutual interdependency, are real though subjectively apprehended, so that, for example, causal relations in nature are perceived within the context of the natural scientists' shared belief in the inanimate nature of material objects, their regularity and predictability. The laws describing how inanimate objects interact are unchanging precisely because the objects involved are inanimate. The laws of interaction of animate, rational and creative beings are much more complex: the principles of interaction themselves may be transformed in the course of social change and the social scientist whose subject matter they are is called upon to assess the ever-changing patterns of motivation, values and objectives, and to discover the effects on the rest of society of the behaviour of each of its parts. Social scientists can achieve none of this in a value-free manner because they will be required to apply a framework of belief in order to be able to interpret the data with which they are presented. They will tend to see manifested those motivations they are most familiar with and prefer to apprehend causal relations which fit or reinforce a pre-conceived picture of reality. This is not because social scientists are consciously trying to distort the facts or to seek out the evidence which supports their opinions. On the contrary, they may sincerely and conscientiously attempt to understand the facts without bias, but it is precisely this which I maintain is impossible – at least in the most widely accepted sense of the word 'bias'. To understand without bias in the way the value-neutral scientist intends is to attempt to understand without the perspective of a knowing mind, and in particular, without acknowledging the role of imaginative involvement in the interpretation of human motives and responses.

The ways in which the elements of society interact are not directly observable. Many things may occur each of which *could* be argued to be the cause for the particular phenomena under investigation. The causal connection itself cannot be observed 'in the act' of producing the phenomena so it is really a matter of considering the alternatives within the context of a whole picture of the ways in which societal elements interact. The overall picture must, at least on the face of it, be relatively consistent and plausible, though however 'matter-of-fact' it appears to be, it will to some extent rely on the knower's personal interpretation of complex sets of relationships. No knowledge of society would be possible without some such picture (or framework) within which further data may be organised, but, as already indicated, this framework will depend upon the perspective of the individual social scientist. An interpretative framework[2] depends ultimately on what the individual chooses to believe, and s/he may or may not be disposed to believe what is actually the case.

A distorted framework based on the individual's misconstrual of causal relationships is (as dissension among economists shows) practically very difficult to overthrow and particularly so since any challenge made at the level of the perceiver's interpretative framework is ruled out of court. The challenge is usually said to have become a political or moral one and therefore not the concern of the empirical social scientist. Within the confines of an allegedly value-neutral social science, one cannot call upon a person's sympathies or ask someone to alter their basic perspective so as to perceive social/economic interaction in a new light. All one can do is challenge the consistency of a position on its own terms, (ie. within a *given* interpretative framework), or, in conjunction with the presentation of one's own account, point to a set of facts and figures which tend to support this account over others. This ultimately convinces no-one and in debates of this kind one finds oneself continually running up against the other person's favoured facts, supporting their own favoured appraisal of reality. There can be no meeting point and therefore no resolution.

Within the confines of the value-neutral methodology we see such clashes of opinions which can never be resolved into agreement because the means for attaining agreement (that is, for radically altering another person's perspective) are out-lawed as 'unscientific'. When all that can be appealed to is positive empirical data then insufficient grounds are made available for affecting the individual's interpretation of essential inter-relations. Value-neutral argument has not the power to re-shape people's conceptions to any great extent. It can encourage greater consistency within a single world-conception, or stimulate the systematic search for, and ordering of data along certain lines (which *are* positive features) but it cannot revolutionise thinking and this is essentially what is needed to procure a more widespread agreement in social scientific research.

Natural scientists rarely need to argue at the level of the interpretative framework to procure agreement. Due to the less unstable, inanimate nature of their subject matter[3] they already possess a more or less common framework for interpretation, though it might not be the best one, objectively speaking. (On this point see the previous chapter on the natural scientists' drive for domination and control of the natural world.) Moral and political standpoints do not so obviously hinge upon the rival theories of physics or chemistry, whilst in social science rival theories are very commonly linked to specific political or moral standpoints.

Let us take as an illustration of seemingly irresolvable conflicts in social science the academic disputes between neo-Keynesian and neo-classical economists over the status of the theory of distribution. The former group tend to argue that once either the wage-rate or the rate of profit is known so too are prices. Thus, conceptually, the theory of the distribution of income between wages and profits precedes the theory of value. For the neo-classicals, however, the theory of distribution is simply an aspect of the marginal theory of value. Obviously linked to this academic clash of opinion are deep-seated ideological differences concerning the functioning of the capitalist system itself. The neo-Marxist/neo-Keynesians see capitalist institutions, such as private property, and the existence of wage-earning and entrepreneurial classes, as giving rise to conflicts between the classes. It is argued that the distribution of national income can therefore not be understood independently of the institutional, conflict-torn nature of capitalism. Neo-classicals see this same economic system as a rather complex, impersonal mechanism, running smoothly for the most part, or at least in principle capable of running smoothly if the mechanism is properly understood and its parts adjusted or maintained. Essentially then the neo-classical picture is one which suggests harmony (or potential harmony) between the various socio-economic groups.

Regarding these two schools of thought, I do not doubt that some very cogent arguments could be ranked on each side, and that controversies which have arisen between the sides have resulted in greater consistency and clarification *within* each school of thought. What I very much doubt, however, is that the debate will have produced many converts. One who stands firmly in the neo-classical camp will not go renegade to join the neo-Keynesians (or vice versa), or if someone does, it will most certainly not be the result of their having made some further observation which had eluded that person previously. More likely it will be the result of a person's adopting a *new* interpretation of *old* facts.

G.G.Harcourt remarks in *Controversies in the Cambridge Theory of Capital*[4] that in the late 1960s if you knew which side of the capital debate an economist was on then you could be fairly certain how s/he felt about the Vietnam War. That it is at all possible to make such inferences suggests that in challenging a position in economics one is challenging much more than the logical consistency or factual accuracy of the theories: a whole belief-system, or framework for understanding, is at stake which, if it is seriously undermined, will have repercussions throughout one's life. To return to our example above, a successful challenge might mean the overthrow of that disposition which inclines one to see people as interacting more or less harmoniously, with the related attitude that one is more or less satisfied with the status quo. On the other hand, it may mean the disruption of an attitude which inclines one to see conflict and injustice everywhere and which encourages the development of an attitude of discontent or of rebelliousness. Peculiarly, people of both persuasions live in the same society, even in the same town, mixing in similar social circles, but each is predisposed to interpret their experiences in a certain way and empirical data alone will not determine whether it is the person who perceives harmonious activity or the person who perceives inequality who is nearer to grasping the true nature of social interaction. Neither harmonious nor exploitative modes of interaction are amenable to direct observation.

The difficulties in determining the correct way to construe empirical data would seem to suggest that we should remain open to the various explanatory possibilities. Instead, what has happened is that the requirement of value-neutrality, ruling out far-reaching explorations of rival positions, has limited the possibilities for rational comparison and evaluation. This makes the resolution of long-standing disputes extremely unlikely.

Social scientists can only hope to resolve their disputes by actively involving themselves in discussions which span beyond the restrictions of current scientific debate, acknowledging the role that different views about society and social objectives play in shaping rival social scientific theories. To involve oneself in such a discussion is to enter upon the consideration of one's most basic values and for this discussion to be rational, and ultimately conclusive, it is necessary that the participants be willing to exercise their capacity for full imaginative involvement with the object of enquiry. Only in the exercise of this capacity are the *actual* relations of interdependency, (of harmony in interaction, of domination or suppression, etc.) likely to be apprehended. It is 'reality-encompassing' rationality (referred to earlier) which allows us to perceive complex sets of inter-relations as they really are, and ideological distortions[5] which cause us to re-interpret relationships as other than they are. Disputes which are apparently factual (amenable to value-neutral study) are, on closer inspection, only capable of resolution when the full scope of human rationality is admitted as legitimate and the discussion is opened up into one that is essentially value-laden. Social scientists who limit the scope of their investigations by demanding value-neutrality are making comprehension of social reality impossible because their methodological assumptions are based on an alienating, impersonal view of their distinctively human object of study. It is of no surprise that only limited agreement has been achieved within the social sciences for without the touchstone of reality there can be no unified body of opinion in *any* field of academic enquiry.

The alienation of the social scientist *as an individual* has a number of related aspects: not only is s/he declining to get involved with reality fully by demanding the value-neutrality of all enquiries, s/he also (often explicitly) refuses to bear responsibility for the consequences of decisions made upon the basis of research. The social scientist will usually be content to say that his/her role is one of discovering and presenting 'value-free' factual data to those whose responsibility it is to make policy decisions on an informed basis. (And it is perhaps also convenient for the social scientist to absolve him/herself of responsibility when the consequences of policy decisions prove distasteful.)

What this attitude indicates is a tendency, widespread in modern society, to section off areas of one's life or, more broadly, sections of the community, for certain kinds of specialised activity which, viewed in themselves, are alienating forms of activity because they require individuals to withhold, or to limit, the use of their rational faculties. Thus in one sphere of one's life (e.g. the working sphere) one may be required to apply procedures, dealing with numerous individuals (or 'cases') in an uninvolved manner and according to a set of rules which are designed to make the process uniform and impersonal. In other areas of one's life (e.g. family life) quite the opposite may be required: individual members of the family have specific needs, and successful or praise-worthy behaviour in *this* sphere is thought to consist in being sensitive to the individual's needs and in being able to adapt to new circumstances rather than to merely apply some hard and fast rules about, for instance, child-rearing.

For the social scientist, the life-division consists in maintaining impartiality through the adoption of a supposedly value-free methodology. Where, in other contexts, the individual is accustomed to observing, assessing, and then acting upon the facts as s/he perceives them, in the work context s/he must be content to merely observe, assess, and present the facts for others to act upon. (And the social scientist will often be prepared to accept another's decisions provided they are acknowledged as the 'legitimate' authority.) Clearly, what is going on is that the academic is allowing him/herself to become instrumental in the achievement of the aims of others. Researchers cannot know exactly how those in authority will respond to the data they are given, and although only instrumental rather than decisive in the achievement of government aims, they cannot rightly absolve themselves of responsibility for actions based on their apparently factual accounts. Suppose a man tells a child how to load, aim and fire a gun but says nothing by way of a recommendation that the child *should* fire the gun. I hardly think the man escapes our condemnation if the child then goes out and shoots somebody.

It might be thought to justify or explain the incompleteness of the social scientist's rational activity by pointing out that the data they collect and analyse does not relate to any specific individual's life and proposed activities, but to the life of the community as a whole, and that the factual picture which the social scientist develops is one intended to guide the actions and objectives of the community. As one individual, it is not for the humble social scientist to set community objectives, although in his/her role as a member of the electorate s/he may play a small part (voting once every five years) in choosing the community's leaders who will then select ends and means on everyone's behalf. Thus it might be conceded that the rational activity of social scientists *qua* social scientists is incomplete: it has no direct outcome in terms of individual action but that is only because of the peculiar nature of their work-activity, which is concerned with the study of large-scale social phenomena rather than with what lies within any individual's immediate experience. Admittedly, perhaps, this represents an unsatisfactory situation for the social scientists, and it is not surprising that many have felt compelled to act upon their academic findings, to unify their lives, translating theories into practice by entering into public life as politicians, community leaders or social reformers. Those who go to such lengths may be commended for the attempt at unifying their theoretical knowledge with practical action, but such attempts are not in themselves sufficient to overcome the destructive effects of a subject-limiting conception of rationality.

Fully rational behaviour requires that the individual act autonomously on the basis of whatever knowledge s/he has acquired for herself and not on the basis of false knowledge claims accepted on the authority of others. Such an individual will have selected his or her own values through the exercise of rational faculties which allow the apprehension of actual relations of interdependency, of exploitation, domination, harmonious interaction and so on, and which will have involved the exercise of imaginative capacities and sympathies. In addition, the exercise of what I have called 'encompassing rationality' will suggest to the individual proper goals or objectives since such things are only given in the awareness of the inter-relations which are affected by purposive activity. (I have already argued that objectives are not supplied by the perception of physical objects. This is why positivism cannot consistently offer us any objectives or support purposeful activity.) What the possession of encompassing rationality *cannot* be made consistent with is the systematic interference with individual autonomy as represented by the various activities of government. There may be occasions when an intervention, to prevent an individual acting autonomously, *is* called for (for instance, to protect the life of another threatened by the proposed action) but this is a matter for the intervening person to decide; it is a matter for conscience and cannot be settled in advance, by a non-specific decree passed and imposed by others. If we turn constantly to some pre-determined set of rules, looking for guidance in action, we are suspending our own, critical, decision-making faculties; we are looking *to others* rather than choosing to get involved with reality for ourselves. The result is that we loose the ability to adapt and to think on our feet; we no longer know *why* we must do a thing, only *that* we must do it. In such a condition, we are not autonomous: we have become like automata, with our rational, moral decision-making faculties worn away through institutionalised neglect.

None of us can be fully rational, and therefore sympathetically involved with the needs of others to lead a fulfilling life, and at the same time wish to take up a significant role in shaping the lives, and options, of others, claiming an authority that, by its very nature, places rational constraints upon all those who choose to acknowledge it. The desire to exercise an influence above that which the common man has over his fellows, springs from the alienated condition of the politically ambitious individual. It reflects a separation from reality and a misguided desire to overcome this separation by means of domination, that is, by imposing one's 'blue-print' for society on others at the expense of individual autonomy and, not infrequently, of material well-being.

To return now to the social scientist who seeks to unify his/her life by entering the public domain. Where this activity involves someone in the imposition of objectives, inspired perhaps by academic research, on the community as a whole then the individual nevertheless betrays an alienated disposition and displays the limitation of the use of his/her rational faculties. This criticism will apply to economists or political scientists who concern themselves with the study of macro-economic objectives or with the various forms of political organisation and control. Whether or not they carry through their implicit recommendations into political life, such social scientists are involving themselves in intellectual activity which is part and parcel of an alienated way of thinking: it involves the unexamined presupposition that large-scale social/economic manipulation by a minority is rational and expedient, and also the presupposition that what is merely expedient can properly be called rational.

The institutional and macro-economic framework of society is essentially authoritarian, or at any rate it is characterised by a real imbalance of power between socio-economic groups who may, among other things, be classified according to how much power, and the nature of the power they possess. By 'authoritarian' I mean that the decision structure of society may be observed to have a pyramidic form so that decisions made at the top of the pyramid affect not only the action of those responsible for the decision, but also the lives of greater numbers of people beneath them in the structure. Those affected are not themselves acting upon the directives of their own rational faculties but have directives imposed upon them if the pyramidic decision structure is backed up by powers of enforcement or by more subtle, non-rational forms of persuasion.

The social scientist involved in the study of such structures, whether they be essentially economic or overtly political, is serving to reinforce the conditions which have undermined the rational faculties and autonomy of so many, for so long. In impartially presenting the 'facts' about the ways in which economic, social and political institutions interact they are, as indeed they claim, providing the basis for those with authority to make informed decisions and to translate their objectives into effective action. Allegedly impartial social scientists cannot, however, excuse themselves of responsibility for their actions any more than can the father who tells his child about the workings of a gun. The nature of the factual account given will itself suggest a course of action because no account of social interaction would have been possible without some assessment of the network of causal inter-relations, based on (in almost all cases, *pro*-authoritarian) evaluative presuppositions and beliefs. If we approached our efforts at understanding society *without* the implicit authoritarian assumption, but merely with a view to understanding the broader context for our *own* activities and choices, then our theoretical explanatory structures might begin to look very different.

The key point here is this: the whole way in which social scientific activity is conceived is determined by the assumption that research is meant to inform those who must take decisions on society's behalf. *Whatever* one thinks about the role of government, it should be apparent that this assumption is grounded in a pro-authoritarian, anti-libertarian political philosophy and, as such, is not itself value-neutral. The 'value-neutral' conception of social scientific methodology is not obvious or uncontroversial from all possible political perspectives: it is a methodology which seems natural *only* from the point of view of (what is undoubtedly) the dominant political ideology.

Social scientists probably prefer to think of themselves as above ideology, or at least as capable of rising above it if they strive towards the ideal of value-neutrality. They are misguided. To fail to acknowledge their factual, 'value-neutral' accounts of society as built upon, and implying, certain values is an act of self-deception, but to go further and to present the 'value-neutral' account to others, knowing that it will direct their action, is sheer irresponsibility indicative of a limited use of their rational faculties. The social scientist's co-operation in the process of man's domination of man is merely the counter-part in the social sphere of the natural scientist's contribution to man's domination of nature.

2. The Denial of Objective Value in Social Science

In the previous chapter, I considered the impact on the objectives of natural scientists of adopting an alienating conception of rational enquiry. Although often explicitly denying the objectivity of values, natural scientists have nevertheless selected a more or less commonly shared goal which, in turn, is based upon a commonly-held, though impoverished, set of values. In the social sciences we may also observe the adoption of a generally unchallenged set of values which have much in common with those of the natural scientist because they spring from the same source. In a world which is seen as lacking objective value, the only available motivation seems to be individual, or species, self-interest; in place of material domination the social scientists substitute the concern for ever greater material acquisition; and, instead of controlling and stabilising nature, social scientists are committed to controlling and stabilising human societies, regardless of the effects on that which they (or others) seek to control.

The uniting themes which run from natural science to social science, through to everyday life, are those of egoism and the drive for power. Egoism arises because our condition of alienation makes it difficult for our sympathies to pass beyond ourselves, and power is sought because in domination or possession the external object is crudely united with its master. The social, like the natural scientist, rarely stops to ask why we should accept these bases for action rather than any other. Since the scientist rejects the objectivity of all values why then choose to value the furtherance of even one's *own* interests?[6]

In Weisskopf's discussion of alienation and positive economics he describes the discipline of economics as 'value-empty', his choice of the word 'empty' implying a criticism which the term 'value-neutral' wards off.

'Neutral' suggests impartiality whereas 'empty' implies impoverishment and although I agree that economics (in common with all positivist social science) is impoverished by the adoption of the value-freedom doctrine, I do not think it is strictly true to say that it is value-*empty*: the fact that values are concealed and their presence explicitly denied does not imply that they are genuinely absent. Even positive social science is action-orientated, or at least is capable of offering suggestions for action, and as such it must be a form of enquiry into the network of causal relationships which exist between people, institutions, governments and economies. Unavoidably, the manner in which these relations are apprehended is distorted by the position which the enquirer has adopted towards the object of enquiry: the enquirer sees him/herself as set apart from social reality, observing it with indifference so that s/he might discover the most effective means of social or economic control. This perspective infiltrates the study not only at the level of determining objectives or areas of interest, but at the more profound level of influencing the interpretation of data. The social scientist can never achieve an 'indifferent' interpretation of what s/he might call 'raw facts' because the act of understanding itself involves more than a passive perception of empirical data which is imposed, unsolicited, upon the senses. The social scientist, to understand anything, must get involved with the object of interest; s/he must perceive it within a context, in its relation to other things. It is this context and the perceived relations between people and things, or social structures, which are most profoundly affected by the social scientists' alienated outlook. In particular, the apprehension of causal relations which exist between institutions and individuals – relations whose exact nature cannot be demonstrated conclusively because they are only subjectively grasped – will be entirely shaped by the ideological disposition of the allegedly 'neutral' observer.[7] Thus we see in economics that there is still little agreement about the interpretation of crucial statistical data. Monetarists' and Keynesians' views on managing the economy differ so greatly because of their different views on the causal relationship between changes in the rate of interest, the money supply, and savings and investment. Keynesians generally believe that saving and investment are interest inelastic and for this reason see the manipulation of aggregate demand as the more effective tool in achieving their chosen macroeconomic ends. Monetarists, on the other hand, see saving and investment as comparatively responsive to changes in the rate of interest and for this reason advocate policies of money supply management. The 'raw data' available will not settle this academic and practical dispute either way, although both sides will persist in believing that they are involved in 'positive economics' and that ultimately such questions can be settled by a correct and truly impartial consideration of the facts. Economists perceive themselves as being involved in a value-neutral study but in reality their study of social and economic inter-relations can only be value-laden, for it is their values and beliefs which provide the context in which causal relationships are subjectively apprehended. If economists are mistaken or uncritical in their choice of values, thinking these do not concern them in their capacity as social scientists, then the result will be a distorted view of relations and inter-dependencies which exist in the real world.

It would be wrong to describe economics, or any other so-called value-neutral social science, as value-*empty* because if value-emptiness were possible such studies would provide nothing that we could recognise as knowledge and they would certainly imply no course of action for either governments or individuals. This is because (as argued in Chapter 3) action requires possessing some view on the nature of our relationship to the world, and this possibility in turn suggests the application of a value-laden framework for interpretation. If economic theory *could* be value-neutral, if a theory could evolve without reference to the totality of someone's belief system, it would be hard to see how such a structure could have any implications for policy. Theories developed from the perspective of no-one in particular could surely not be useful to anyone in particular.

It is a peculiarity more often noted amongst social scientists how little agreement over the facts has been achieved. Although both natural and social scientists are involved with interpreting data within a framework of beliefs, the latter group are concerned primarily with *living* subject matter, that is, with active individuals and the complex set of relations which they create in forming and re-forming society. Thus by comparison natural scientists appear to have an easier task – their subject matter does not shift and change. Physical material does not interact with itself to create new laws as the basis for future interaction. On the whole, the laws of nature appear as fairly constant and once we have discovered them (if we really have discovered them) then we feel we know them for good. The scientist does not expect or need fear that the laws of nature will change tomorrow, but this is precisely the thing which the social scientist must fear – at least if s/he is alert and aware of the peculiar nature of the subject matter being studied. Living things are essentially active and, more significantly, they are creative. This means that we can act purposively to re-shape relations between people, and also the balance of power between nature and humanity. Unlike matter, we are autonomous and can create over and again new principles for interaction. The way in which we do this and the principles for interaction which we knowingly or unknowingly choose, depend upon what beliefs we hold because it is these beliefs which provide the framework for our interpretations, which determine what we are disposed to apprehend. Interpretations, particularly those of academic social scientists whose opinions are respected and often highly influential, may serve to reinforce current value attitudes if they are presented uncritically as value-neutral facts. This was the criticism Marx made of the classical economists, that they presented their theories and explanations as if they were eternal truths about the interaction of human beings in an economy when, in fact, their theories applied only to a certain kind of economic organisation, that of capitalism.

We cannot free ourselves from all beliefs and values without ceasing to be active, rational beings but *as* rational beings we can hope to free ourselves from the grip of irrational beliefs or ideologies. Indeed this is one of the aims of rational mental activity. Social scientists must concern themselves with uncovering ideologies (they need to do this to understand how society, or the economy, is actually working) and then examine their findings critically with the purpose of discovering new, more fully rational principles of social and economic interaction. Ultimately the question of how we should move forward, as individuals or as a social group, is the one which gives sense to the search for knowledge in any sphere of human enquiry.

I have referred already to the peculiarities of the social scientists' self-transforming subject matter, human society, and to the role which belief-systems (some of which may be ideological) play in the interpretation of social relations. It should be clear already that the social scientist faces interpretative problems which the natural scientist does not, and I believe that these problems account for the differences which Kuhn observed between social and natural scientific communities.

Kuhn argues in *The Structure of Scientific Revolutions*[8] that the history of natural science is characterised by the dominance of successive paradigms. Each lays down a set of practices and assumptions as well as a common methodology within which scientists define and solve their problems. A scientific revolution occurs when this once widely accepted framework is challenged by another which directly contradicts it or is said to be

74

'incommensurable' with it. The old ideas and procedure for enquiry are then overthrown and a new paradigm is established. Kuhn remarks however[9] that no such pattern can be observed within the social sciences and puts this down to the relative newness of such disciplines. This leaves us with the implication that when our social sciences 'mature' they too will be characterised by a high level of agreement on fundamental facts and principles. I do not believe Kuhn's is a realistic view.

I commented earlier that natural scientists rarely need to argue at the level of the interpretative framework to procure agreement and I attributed this fact to the less unstable, inanimate nature of their subject matter. In contrast, the subject matter of the human sciences is by no means subject to eternal, unchanging laws of interaction but is at once both self-comprehending and self-transforming. Thus, any study of human society will always lean more heavily on the researcher's own moral and political viewpoints (views about what we are and what we *should* be) to achieve a coherent interpretation of the data. It is more apparent in social science that the commitment to some particular theory is intimately bound up with commitment to, for example, a political position, such that a challenge to a theory is seen as a deeply personal challenge with far more at stake for the defender than the theory itself. Such links between theories and general world outlooks are far less easy to detect in the natural sciences, but this is not to say they are *not* there. In Kuhn's account of the processes at work during a period of scientific revolution, he identifies as a characteristic of the revolutionary transition, a marked lack of agreement over fundamental assumptions and practices. In essence, what is occurring in such a transition is the re-evaluation of the previously shared interpretative framework and in the course of this re-evaluation scientists will typically often find themselves embroiled in distinctly non-scientific discussions. So, for instance, Kuhn mentions metaphysical differences of opinion, disagreements over the role of science in society and over what problems need most urgently to be addressed – all these elements play a part in establishing a new framework or, for Kuhn, a new 'paradigm', for guiding scientists' research activities *after* the revolutionary transition. This means that even once a period of 'normal science' has been re-established, whatever new theories are favoured owe their acceptance, at least in part, to *non*-scientific elements of the scientists' belief systems.

Kuhn concludes as a result of his recognition of non-scientific elements in scientific debate that the revolutionary transition must therefore involve a non-rational choice in favour of the emerging paradigm. He is assuming that social, moral and metaphysical questions which, it seems, have an input into science cannot be debated rationally and thus he is reluctantly compelled to embrace relativism and the idea that real progress in science cannot be demonstrated.

What we see in the work of Thomas Kuhn is the impact on what was once thought to be the most secure and clearly *objective* area of enquiry of an overly limiting account of what the rational subject can accomplish.

In a comparison with the social sciences, the natural scientific community at least seems more successful in achieving value-neutrality. However, on closer inspection, the difference should perhaps be seen as more a matter of degree than of kind. The natural scientist inevitably escapes, for the most part, the worst ravages of ideology whilst social scientists remain inescapably caught up in the full complexity of their individual belief systems, affecting the way in which they interpret almost every minute detail of social reality. The superiority of the natural sciences with respect to the measure of agreement on fundamentals is, however, merely relative; such superiority does not imply that natural science is, in contrast to social science, entirely liberated from the influence of ideology. The detached, essentially alienating manner in which research is undertaken (influenced by the dominant conception of objectivity) has led to scientists viewing the natural world almost exclusively as an object for human domination. Broader agreement is only possible among scientists despite ideological influence because the subject of their investigations operates according to unchanging laws and because the apprehension of cause/effect relationships between physical objects does not lean so heavily upon the observer's interpretative faculties. The uncritical social scientist, unlike the equally uncritical natural scientist, may be dogged at every step by the distortions of ideology and can only overcome them by acknowledging the role which the totality of a person's beliefs and values play in understanding the world.

The value-neutral social scientist, in denying that values play an ineliminable role in shaping and consequently in understanding society, is refusing to face head-on the fundamental cause of the disagreements which are rife within this field of study. S/he has emulated the natural scientist's value-neutral approach, admiring it for its impartiality rather than recognising its impoverishment, but whereas the natural scientist has more or less 'got away with it', the social scientist, whose subject matter is autonomous and self-transforming, has suffered the amplified effects of the distortions of the value-neutral approach. Social scientists have done themselves a great disservice in choosing natural scientists as mentors, and instead of priding themselves on how like the methods of the scientist are those of the social scientist, they would have done better to concentrate on the differences, namely, on how little agreement has been achieved on matters of fact.

It is the purpose of the natural scientist to study the way the natural world *is*, to discover the unchanging laws of its operation, but for the social scientist the purpose differs because of the fundamentally different nature of the object of study. The object of study – human society – is constantly undergoing change, not merely of the kind undergone in nature which proceeds according to unchanging laws, but change according to laws of social interaction which are themselves capable of being transformed. The purpose of the social scientist should be to discover not only what are, but also what should be, the laws of interaction. The social scientist cannot step outside of the creative process of self-transformation because to do so would be an irrational denial of the very purpose of rational enquiry.

At the moment, social scientists under the influence of the value-freedom doctrine are straining to ensure that they observe with rigour the requirements of impartiality, that they observe only the facts and resist the temptation to make evaluations (unless, of course, these are presented as asides and coupled with an apology). The ideological distortions which so constantly plague their studies, the unacknowledged evaluations which are built into their descriptions of facts, must be brought out into the open and critically examined. The way in which they finally portray social and economic reality will unavoidably have an impact on public awareness and cause questions to be raised about future objectives. Social scientists must therefore acknowledge their responsibility in forming opinions, in reinforcing or re-shaping beliefs, and must respond to this responsibility by actively involving themselves in the discussion of what our objectives should be. Until they explicitly acknowledge their part in this discussion, the disagreements which plague social scientists will not be resolved and progress will be curtailed.

Notes

1. Much more will be said on the nature of ideology generally, and on the specific ideology at work in Western society, in Chapters 7 and 8 respectively.
2. An 'interpretative framework' refers to value attitudes and beliefs already formed, and which are sufficiently entrenched to influence the ways in which further experiences become assimilated. A distorted, or restrictive framework for interpretation limits our ability to adapt a belief system to achieve a better fit with reality.
3. I am thinking primarily of subjects such as physics, chemistry and astronomy, though of course biology is also classed as a natural science. The subject matter of biology is not inanimate. However, when studying something as a biological entity (and not as a *social* entity for example) the biological scientist is not generally concerned with those functions which fall under conscious control. In so far as biologists do take an interest in animal behaviour, to that extent their discipline is seen to overlap with others whose scientific status is arguably less clear. The ways in which we interpret behaviour (both in humans and in other animals) seem to depend much more obviously on rival, heavily value-laden positions.
4. Harcourt, GC., *Controversies in the Cambridge Theory of Capital*, Cambridge University Press, 1972.
5. 'Ideological distortion' refers to the distortions which occur when a restrictive framework for interpretation is adhered to. An individual may, in effect, choose *not* to be involved with certain aspects of reality and thus prefer the ideological interpretative framework which serves precisely to keep out those unwanted aspects.
6. Logically, the fact that I am the subject of my own actions does not imply that I must be, or even can be, the sole object of my own concerns. The relationship between myself as a subject of actions and myself as an object in the world, with interests that somehow 'ought' to be recognised by me as a subject, is highly problematic. (On this point, see Parfit, *Reasons and Persons*. Also, Nagel, *The Possibility of Altruism* and *The View from Nowhere*.)
7. I identify the 'value-neutral' conception of science (and of objectivity) as an integral part of a particular ideology which will be discussed in Chapter 8.
8. Kuhn, TS., *The Structure of Scientific Revolutions*, University of Chicago Press, Chicago, second edition, 1970.
9. Kuhn's comments on the lack of an identifiable paradigm in many of the social sciences are to be found in *Criticism and the Growth of Knowledge*, ed. Lakatos and Musgrave, the paper entitled 'Reflections on my critics' §3.

5 Repercussions of a Science-Dominated Conception of Reason

1. The Overthrow of Restrictive Frameworks

I have already referred to the impossibility within the confines of a value-neutral methodology of radically altering another person's perspective because what is needed – the overthrow of an existing interpretative framework – is ruled out as requiring 'non-scientific' arguments. As a consequence of this restriction, widespread agreement is difficult to achieve: a conflicting interpretative framework cannot be touched by arguments which deal only with empirical phenomena. I shall now turn to the question of what sorts of argument *are* capable of overturning the framework within which people interpret their social and economic environment.

In order to apprehend the full range and nature of the inter-relations that exist between objects, social constructs and individuals, it is necessary for the rational agent to become directly involved with the objects of knowledge. This view of knowledge as direct involvement (based on a form of naïve realism) stands in contrast to the view of knowledge which sets the knowing individual at a distance from the world so that s/he does not know reality directly but can only reconstruct it out of epistemological building blocks such as sense-data. On the direct realist account, knowledge is attained when the objective world is correctly apprehended by the knowing subject so that what is objective and independent becomes internalised and subjective. If I may reverse Wittgenstein's famous dictum, when 'my world is *the* world' I may properly be said 'to know'. Following from this, arguments may be said to be rational and purposive when they are capable of persuading the listener to accept what is actual (ie. objectively true) as a part of his or her subjective apprehension. But it is not merely rigorous and systematic argument which is capable of achieving this – often the skilful presentation of an alternative picture of reality, or of reality as seen from another's perspective, will excite the imagination and extend the boundaries of a previously limited or distorted view-point. When this happens, the other person's perspective on reality has *become* one's own, perhaps enriching one's own view, or even over-turning it. When the latter occurs, the insight produces something akin to a conversion, after which one's life and objectives may alter beyond recognition.

The crucial feature of the 'persuasive', in contrast to the 'systematic', form of argument is its capacity to stimulate the listener into emotional involvement, and even to an identification, with the needs, concerns and objectives of a subject who previously has seemed distant and incomprehensible. In seeking to understand human behaviour and society such identification, or imaginative involvement, with what is external to oneself is an essential element in the attainment of a broader perspective. This, in turn, will supply a fuller picture of the variety of human responses, motivations, needs, and perhaps most importantly, of the extent of the suffering unnecessarily inflicted as a consequence of our failures to exercise the imagination and sympathies. In short, what the stimulation of our faculties for active involvement provides is greater contact with the complex whole of reality, including what is perceived by the senses and what is *not* perceived but only imaginatively apprehended, such as relations of dependency, subjugation, mutual inspiration, exploitation, etc., which exist between individuals, and between individuals and objects. An argument that is rational in the limited sense, proceeding from premises that the listener (or reader) may explicitly acknowledge and moving by sound steps to conclusions that the listener must also accept will, often lamentably, have little real effect on the allegedly convinced individual so that we are disinclined to say that s/he really 'knows' for him/herself the thing that has been proved. At best the person could be said to have a kind of 'knowledge without conviction', but I have argued elsewhere that this is not knowledge at all but rather an uncritical acceptance of some fact in accordance with respected authorities or, in this case, in response to the pressure to *appear* rational to the extent that one will nominally accept a conclusion that is logically argued. It is illustrative of the way in which analytical thinking has been divorced from a broader concept of what it is to think rationally, that one can accept a conclusion argued in a systematic, logical manner and yet fail to internalise the knowledge so obtained such that one would wish to guide one's actions by it. An example of this sort of separation of the reasoning faculties from what it is that guides action is the acceptance by some individuals (among them philosophers) of the arguments for animal rights. An alarming number of people will, after 'rational' consideration, acknowledge with apparent sincerity that they really ought not to eat meat or wear leather, but nevertheless *will* go on eating meat and wearing whatever they like. Similarly, one will find many more people who are prepared to accept arguments to this effect:

(a) All people are entitled to sufficient food and medical care;

(b) Whenever I can I should give money to help alleviate the problems of starvation and disease;

(c) I have more money than I need therefore I should give something now to alleviate starvation and disease.

This may be accepted as a good argument but it is by no means the case that all those who formally accept it, act upon it.[1] It would appear that rational arguments, at least of the above kind, are insufficient to really persuade someone to admit what is objectively real into the world of their subjective experience, that is, to make the objective world *their* world in the act of knowing it as it is. To return to the above two examples, we will often find that people not convinced by the arguments alone *will* be convinced by the shocking experience of seeing animals slaughtered or by visiting a famine-stricken country. Such experiences force the imagination out of its slumber so that a direct involvement with, or knowledge of, what is real (real needs, real pain) is almost impossible to avoid. It is as if argument on its own is sometimes incapable of touching upon the reality with which it is apparently concerned. Formally, therefore, one can accept the arguments but still know nothing concrete which would inspire action one way or another. Yet if what one knows is *not* concrete, can we really be said to 'know' anything at all? We can only know what is real, or actual, so if what we claim to know is merely formal, implying action but not inspiring it, what we 'know' is a world equally formal or lacking in substance. It is a dream world which roughly corresponds to the actual but in which it matters not what we do. If what we think is reality does not bear down upon us and demand our involvement then what we claim to 'know' is not reality at all.

Quite often a good argument is taken to be one which is clear, consistent, and made up of factually accurate premises. Attempts to excite the emotions are ruled off-side, or at least discounted as part of the main thrust of the argument on the grounds that they are 'merely' rhetorical – colourful and effective but not strictly rational. What remains of the properly rational part of the argument is intended to elicit an unimpassioned response in favour of the proffered conclusion, and if such a response *is* obtained we are said to have communicated an item of knowledge. What is overlooked here is that, given the view of knowledge as involvement, and knowledge of other individuals as requiring sympathy and imagination, it is *on the contrary* quite proper to excite the sympathies by the use of emotive language or by the vivid depiction of aspects of reality which are not within the experience of the listener. Such methods *are* designed to sway the emotions of one's audience but it is not necessarily wrong to do so; it is not wrong in principle as a method of conveying concrete knowledge, though like all methods of communication, it may be used to express untruths or to distort the truth.

Speakers at National Front rallies may be heard to use emotive language in an illegitimate way, focusing the crowd's attention on isolated incidents where black immigrants have beaten up white people or taken what are supposed to be white people's jobs. This kind of talk is calculated to have an inflammatory effect upon the disgruntled unemployed who are already inclined to see immigrants as an insidious and potentially aggressive invading force. The argument is illegitimate because it takes isolated facts and interprets them within an inappropriate framework. The audience's sympathetic involvement with reality is thereby limited to just those individuals who, for instance, have suffered aggression at the hands of immigrants. The reason for such attacks is conjectured to be, for instance, the blacks' brutish instinct and desire to take over the country for themselves. The framework for the racist's interpretation of facts may even lead him/her to make up new facts to fit and reinforce the picture. Thus, some particularly absurd racists have actually claimed that black people dominate the top jobs in British industry and hold many key positions in government. In such extreme cases, an inappropriate interpretation of reality has resulted not merely in the mis-use of available data, but in the even more bizarre practice of manufacturing new facts and believing in them as completely as any thing else one claims to know. Such 'facts' help maintain beliefs which are perhaps rather flimsy or unsubstantiated, but which are so central to a person's life that they cannot risk a challenge for fear of too great an emotional upheaval.

The purpose of my example here is to acknowledge the role that emotive language *can* play in contributing to the distortion of reality and, at worst, to the complete departure from reality into fantasy and insanity. However, what I believe is illegitimate about the racist's appeal to the crowd's emotions, is the selectivity of the call for emotional involvement. We are expected to concentrate, for instance, on the white victims of attacks, on white people unable to find work; we are *not* expected to extend our imaginative involvement to the black person who cannot find work because of his/her colour (and indeed we may be asked to discount this as a fact on the grounds that it is probably a piece of 'lefty' propaganda).

It is the selectivity of one's emotional involvement with different elements of society which results in the construction of a distorting framework for understanding. Once constructed, this framework will help to compound the problem of isolation from the *whole* of reality because what does not fall within the explanatory powers of the framework will either not be understood, will be ignored, or put to one side as temporarily (though in fact *permanently*) inexplicable. The framework can then only be amended by arguments which extend the individual's emotional involvement to new aspects of reality which go beyond what s/he has experienced, or is willing to experience. Thus it is not in principle wrong to argue emotively. It is only irrational to appeal to the emotions in a selective fashion, isolating and giving undue weight to certain aspects of reality, but it is quite legitimate (and indeed necessary to procure fuller involvement with reality) to appeal to the sympathies in all areas where sympathetic understanding is a possibility. This includes all parties with needs, interests and desires which can be shared and which ought to be taken into account and weighed against one another in situations where they appear to conflict.

This account may seem similar to other attempts to give a rational basis to moral judgments. It accommodates a commonly-shared intuition that a sound moral judgment has to have the characteristics of impartiality and universal applicability. In Kantian philosophy, this requirement manifests itself in the form of the categorical imperative: 'act only on that maxim through which you can at the same time will that it should become a universal law.' [2] For Hare, (*Freedom and Reason*[3]), moral judgments are characterised by their lack of reference to particulars (i.e. their universality) and the preparedness of the individual making the judgment to accept the consequences for each person whom it affects, after imagining him/herself systematically in the position of each such person. Another account, that of John Rawls in *The Theory of Justice*[4], seeks to secure the impartiality of our moral judgments by introducing the requirement that our principles of justice must be such that we would agree to them behind a 'veil of ignorance'. This ensures that we know nothing of our unique personal interests but only what interests we all share in common as human beings, so that no principle could rationally be accepted which arbitrarily favours one human being over another. Of Rawls's account we may be tempted to ask: why should we continue to accept principles we agreed to in a state of ignorance? What *was* rational to accept in the 'Original Position' is not necessarily rational to accept now that we have more information about our circumstances.[5]

Kant, although acknowledging that sympathetic involvement plays a part in reinforcing our sense of moral duty, does not think that the emotions play a *necessary* part; rather he sees them as dangerous to rely on as a guide to moral action. The basic idea is that the emotions may be 'trained' to act in support of moral action but that what is moral is not grasped with the aid of the emotions (through emotional involvement) but only by the rational will acting in accordance with the categorical imperative.

Finally, Hare thinks that the universality of moral judgments is a logical result of a consistent use of language, but he cannot fill out what is involved in doing the rounds of all the concerned parties to see if one could accept a particular judgment on everyone's behalf. Thus he cannot deal effectively with the racist fanatic who feels s/he *can* universally accept intuitively irrational prescriptions.

Where I feel that the above, and similar accounts, are unsatisfactory is that they all fail to make essential contact with the real objects of moral/immoral behaviour – living beings with interests and needs – who are somehow thought to be irrelevant, or at least not accessible to the rational, knowing mind. It is because of this intuitively odd exclusion of beings with interests and needs from the centre of moral discourse that moral philosophy such as Kant's prove ultimately unconvincing.

It is because of the tendency to define rationality in a restricted fashion that many attempts to bestow rationality upon our moral judgments have failed. Sympathetic involvement is thought to be the source of irrational, and also partial, judgments rather than being seen as a pre-requisite for fuller understanding leading to rational, moral behaviour. The rationality of moral discourse is only secured on a naïve realist account of moral judgment whereby one is directly involved and sympathetic to every party, and naturally concerned for others' welfare as a direct consequence of our emotional involvement. Only on a realist, interactionist account of our knowledge of others can their concerns *become* ours and motivate our actions towards what is therefore morally acceptable behaviour. The reason why impartiality is intuited to be characteristic of all truly moral judgments is because impartiality stems from the fullest emotional involvement with *all* the relevant parties, whereas partiality signifies a correspondingly limited involvement with the parties against whom one is said to be biased. In cases of limited involvement, the person may not be said to be behaving either rationally (because he is ignoring aspects of reality) or morally.

The intellectual prejudice against emotive argument goes back at least to Plato, who maintains that what was real could only be apprehended by the faculty of Reason, and that Reason in conjunction with the Spirit, or Will, must harness the irrational appetites which constantly threaten to corrupt the soul. From within the Socratic epistemological framework, the rhetoricians appear as defenders of irrationality, employing arguments whose only aim seems to be to excite the emotions rather than to establish and convey objective truths. It fails to consider that the excitation of the emotions is, at times, the appropriate response of a rational being involved with certain aspects of reality. It also rules out the idea that the enhancement of the sympathies could produce greater understanding where before there had been none, or little understanding produced by less emotive (more 'rational') arguments.

On the Platonic view, understanding was not the domain of the emotions but purely of Reason, so that Rhetoric alone could have no virtue but as a medium for explication under the command of Reason. Again we see here an artificial division of the human personality into elements with distinct functions, from which it is but a short step for Plato to argue in his social philosophy for an alienating division of the tasks of society: he argues[6] for the formation of an intellectual élite of Guardians whose rational temperament enables them to restrain the appetites, ignoring what merely excites the emotions, and to grasp instead the nature of ultimate reality which lies beyond the grasp of ordinary men. The Guardians' rationality is such, however, that according to Plato (*The Republic*, Book 3) they would believe that the once hard-working carpenter who is too ill to continue earning his living should sooner refuse prolonged medical treatment and die than become a burden to the State. This somewhat callous conclusion ought not to strike us now as coming from one who is particularly in touch with reality but rather from one who is quite hardened to the reality of pain. The emotional numbness of the Guardians does more to distort their view of reality and what sort of thing may be termed 'real', than to enhance it.

By the time of Aristotle, Plato's pupil, we see a limiting conception of rationality, one which has since become the hall-mark of Western culture, decisively established. Rhetoric, which Plato, through Socrates, feels the need to attack as a major threat to objective knowledge, is placed within the Aristotelian system[7] in the category of 'Practical Science', secondary to 'Theoretical Science', in which Aristotle himself is more interested. Thus, emotional involvement and rhetoric are denied any *direct* role in the discovery or communication of important areas of knowledge and the latter, rhetoric, retains what intellectual respectability it now has only by virtue of its usefulness as an ornament, or device, to enhance the bare bones of a rational argument.

Like Plato, Aristotle maintains that ultimate reality is not to be discovered in the world of appearance but rather belongs to the 'essence' of things, not (as Plato believes) to 'Forms', but to the underlying substance in which properties or appearances reside. It is not until Galileo that the Aristotelian conception of knowledge and of ultimate reality is finally overthrown in favour of a conception in which appearances are the object of theoretical scientific knowledge. This represents almost a complete reversal of the epistemology that had gone before but shares in common with it the feature that the ordinary man's knowledge of reality remains undermined. Sociologically, what occurred is that the domination of an educated, aristocratic élite, who had access to the mysterious world of 'essence' beyond the uneducated man's world of appearance, has been superseded by a class of scientific élite with the necessary understanding of, and access to, the instruments of technology. Thus, the scientist has replaced the erudite priest or divine sovereign as the sole possessor of knowledge.

Edmund Husserl in *The Crisis of European Sciences*[8] makes this same observation, that since Galileo and what he calls 'the mathematisation of nature', the ordinary man's knowledge has been devalued. Nature as it is encapsulated in the highly technical, mathematical language of the scientist has usurped the place of nature as 'life-world', or the world of pre-theoretical experience. The result has been the lack of confidence of the ordinary person in his/her capacity for understanding, and a corresponding degree of reverence for the knowledge and skills of the scientist.

One further significant result, for explanatory purposes, of both the Aristotelian and Galilean epistemologies, has been the relative demotion of the humanities as bodies of knowledge: the novel, music and art are expressions of feeling rather than systematic studies of nature (either in its essence or appearance), and whatever knowledge they express is subjective or emotionally-tinged. Roughly speaking then, science is concerned to broaden our objective knowledge, whilst the humanities – in so far as their methods and aims are distinct from the sciences – are purely for our entertainment or, more generously perhaps, for our cultural betterment. To take but one of the forms of art, the novel, a great source of imaginative stimulation leading to true knowledge and rational activity, has been harmfully overlooked by people narrowly employing 'objective' scientific methods. To remedy this, the deeply-entrenched distinctions which have arisen between the sciences, and in particular between the social sciences and humanities, must be broken down so that the pursuit of all forms of knowledge is seen to involve the exercise of essentially the *same* mode of thought, that which is characterised by the reality-encompassing mental act.

The social scientist, to resolve his disputes about the nature of economic or social reality, must extend the range of arguments s/he is prepared to consider to include (for instance) the persuasive arguments of those who attempt to depict and come to grips with social reality through the medium of the novel. The individual may then judge the truthlikeness of the novel (as s/he judges scientific theories) by the fullness of the picture it gives, by the variety of perspectives covered, and, consequently, the honesty with which society is portrayed. It is worthy of mention that it is the good writer, and *not* the good social scientist who is commonly thought to be perceptive and who is renowned for his/her keen powers of observation. These are not skills to be lightly overlooked in studies of human behaviour.

2. The Science and Art of Reasoning

I have tried to argue that there is a role for persuasive or emotive argument in rational discourse but I have perhaps not made it sufficiently clear that I do not wish persuasive argument to replace or usurp other forms of rational communication, so reversing the present state of things. What I am in favour of is a re-uniting of two aspects of rationality which have been made distinct, that aspect which is thought to characterise scientific reasoning, and that more intuitive or imaginative aspect which apparently belongs to the Arts. This idea of a single mode of rational thought permeating all intellectual endeavour is not new: Einstein himself remarked that '...real science and real music demand one and the same thought process.' Since the remarkable developments that have taken place in physics, more and more scientists have begun to acknowledge that the process of gaining knowledge has been far more intuitive than it has been strictly logical and that when new discoveries emerge, far from being systematically reasoned out and decisively confirmed, they take a rather more confused, incomplete form and are not always fully understood by the discoverer him/herself. This admission has been made not only by physicists. The biochemist Albert de Saint D'ierdi has expressed the view:

> research is rarely directed by logic; it is to a large extent guided by hints, guesses and intuition...The basic material of research is imagination, into which are woven threads of judgment, measurement and calculation.

I think this account of the basic research materials could serve just as well as a definition of rationality with all its aspects re-united.

Although conceding an intuitive element in their research, I do not think it necessary for scientists to go further and concede the essential irrationality or illogicality of their methods, for this only follows on a limited conception of rationality and perhaps a similarly limited conception of logical thought.

In *A System of Logic* [9], Mill argues in favour of a broader conception of logic. He says that it may be seen as both the Science and Art of Sound Reasoning and recognises that the term 'reasoning' itself is an ambiguous syllogism, but it may also refer to the mode of inference whereby one. It can be, and often is, used to mean pure deduction as exhibited in the generalisations are derived from particulars, or whereby assertions are offered as evidence, or justification, for the acceptance of other assertions. Mill's own definition of logic takes the broader form which includes the two latter modes of reasoning. It is interesting, however, that Mill's reflections suggest some reservations we might have about even the fuller definition which does not seem to capture *all* that is commonly understood as being logical. In ordinary usage, ideas connected with logic include the precision, perhaps even the persuasiveness of language, and the useful or illuminating classification of facts. We speak sometimes of a 'logical arrangement' of ideas, or of expressions 'logically defined' as much as of the logical deduction of conclusions from premises according to strict and inviolable rules. Mill offers by way of illustration our use of the term 'great logician' which applies equally well to the man who has a great command over premises as a basis for argumentation as it does to the man known for the accuracy of his deductions. When we begin to consider these common-sense usages, it becomes clear that the criteria for what may be termed 'logical' are blurred at the edges with the criteria for what is illuminating, perceptive, persuasive, or simply aesthetically pleasing. The role of logic as it is more broadly conceived, is to provide the guidelines for rational discourse, but these guidelines may take many forms other than those of deductive reasoning.

It is because of Mill's fuller conception of rationality, and of what counts as a logical argument, that he is able to provide some basis for the belief that moral judgments may be rationally justified. Given that justificatory reasoning is valid, we may use as evidence the fact that something produces happiness to support our assertion that it is good and therefore the proper goal of purposive action. Although maintaining the possibility of attaining objective truths on moral issues (determined, he suggests, by an accurate utilitarian calculation) Mill does not maintain the evidently unreasonable position that such truths, or objectively established values, should as a consequence of their truth be placed beyond all further critical assessment. Whereas a successful deductive argument will apparently lift its conclusion above criticism, the conclusions of inductive or justificatory arguments must remain provisional given the possibility of further information coming to light. It is because of the provisional nature of their conclusions that scientists and philosophers have often felt compelled to reject the latter forms of argument as incapable of providing true knowledge – but as I have argued earlier in Chapter 3, they are impressed by a misleading analogy that says knowledge must be built on firm foundations and thus be characterised by a certainty which implies that, once established, its truth cannot be challenged. On this view, however, human activity breaks down, and with that I believe we must abandon this overly-restrictive conception of knowledge.

Mill rightly points out that by far the greatest portion of our knowledge (what is called 'ordinary knowledge') is arrived at by just those forms of reasoning which deductivists despise for their provisional nature. He observes that:

> inference is the only occupation in which the mind never ceases to be engaged...[It is] the logic of Science [as well as] that of business and life.[10]

We might also add here that it is the logic of moral discourse, and that it is Mill's provision for, and defence, of this mode of reasoning which provides a basis for the classical form of liberalism.

3. Restrictive Rationality and the Fake Morality of Modern Liberalism

Basil Mitchell, in *Law, Morality and Religion*[11], gives a useful account of Millean liberalism and contrasts it with a later form of liberalism in which the belief in the possibility of an objective morality, rationally determined, has been abandoned. Owing to Mill's broader concept of rationality, coupled with recognition of the provisional character of inductive or justificatory truths, he is able to maintain the possibility of discovering objective truths without also needing to uphold the infallibility of the discoverer. Thus, the Millean form of liberalism has not given up on the idea of there being a unique set of moral values upon which every rational person could agree but which no-one is compelled to embrace. This lack of compulsion to assent to moral truths stems not only from Mill's acknowledgment of one's potential fallibility but also from his recognition of the value of freedom to contest allegedly established truths. This freedom is required not merely for the individual's own happiness and self-fulfilment but is itself a requirement of productive, rational debate. Mill therefore stresses in *On Liberty*[12] that even if we *were* sure of the falsity of an opinion, we would be wrong to stifle it in any case.

Mitchell contrasts Mill's position with that of Plato who, like Mill, maintains a belief in the objectivity of values but who believes that infallible knowledge of the Good legitimises the Guardians of his *Republic* in their suppression of the critical faculties of the citizen. The Guardians' role is to do the thinking for the rest, and to organise society in accordance with what they (infallibly) perceive to be good. It is unfortunate that from within the framework of modern liberal thought, the maintenance of a belief in objective moral truths is often seen as the first link in a chain of beliefs which leads to the justification of totalitarianism; but as we have seen from the contrast of Mill with Plato, this conclusion need not follow. It is a result which depends also on the belief in one's own infallibility and in the right to impose what one infallibly knows on others. (One might, of course, infallibly know that one has no such right in which case totalitarianism still does not follow.)

Modern liberal thought may be compared and contrasted with Mill's classical liberalism in that it has abandoned the possibility of a rational, common morality, but has salvaged from Mill a (more superficial) respect for individual freedom. Since no morality has any greater claims to rationality than any other, it is the tolerance of diverse moralities which is prized above all else. Tolerance itself, however, could not be consistently claimed as a universally-prized (because *objective*) good; rather it is an attribute which springs from the lack of any good reason to object to another's morality. Freedom is simply the necessary background against which people may go about their business according to whatever principles or values they happen to prefer. In modern liberal society, freedom of choice is something which we are said to 'value' or respect, though strictly speaking, nothing can be valued once the belief in the objectivity of values has been abandoned. The 'value' which is attached to freedom of choice is, in this case, merely the indifferent acceptance of the principle of freedom for lack of any value-based reasons which warrant interference with another person's choices. A choice is therefore valued not because what is chosen is good but simply because it is freely chosen, and there is no possibility of the free individual making wrong or alienated choices because there are no independent rational grounds for assessment.

The highly influential work of John Rawls most notably embodies the value-impoverished philosophy of modern liberalism. In *A Theory of Justice*[13], Rawls derives principles of justice such that individuals may, once a 'veil of ignorance' is lifted, pursue whatever preferences they have, provided that they refrain from interfering with the activities of others. This latter requirement is what they will have consented to 'behind the veil' before they could have known what choices or preferred values they would adopt. It is not clear on Rawls's account why we should abide by what is (allegedly) rational to concede in the artificial environment of the 'Original Position', but it does seem that Rawls has built into his theory the virtues (much cherished by Americans and which he confidently expects everyone else to accept) of liberty and equality *and nothing else besides*. What the 'Original Position' gives us is a device to ensure that whatever principles of action we choose (and we may choose any) we do not act so as to interfere with others. The device secures the impartiality of our judgments without providing any really *positive* reasons why moral judgments should be impartial. If reasons could be provided then I would argue that more follows than is allowed by modern liberal thinkers: they would have to admit the possibility of positively evaluating not only freedom of choice, but the outcome of certain objectively praiseworthy choices.

It is worth contrasting here Mill's attitude to freedom with that of modern liberals such as Rawls. Mill's evaluation of freedom is what I would call 'positive', not because it is valued in itself (which it is not), but because it is valued as an essential part in an objective morality. What is ultimately of value is the greatest happiness of the greatest number and this end is shared by all those who wish to claim their actions as moral. The important point is this: that the moral person wishes happiness not only for him/herself but for all others capable of experiencing pleasure and pain so that when people think morally they are, in fact, passing beyond themselves to actually *share* the interests of others.

Within the Rawlsian system, however, the individual only shares the interests of others insofar as it is in their interests to be free and to have equality of opportunity. The requirements of the 'Original Position' are such that no-one knows what their interests or preferences will be so that it is simply imprudent to consent to partial or inegalitarian principles of justice. One is not required by Rawlsian morality to go beyond oneself to actively value another person's interests; rather we are asked to imagine that we have no preferences, in short, *no* identity, when we choose our principles of social justice. The 'Original Position' is either populated by a lot of characterless, descriptively identical individuals who could only want the same things prior to knowing their full identities, or else it is effectively populated by one imaginary individual who, not surprisingly, will choose to value his/her freedom of choice once the veil is lifted and s/he becomes a multitude of individuals with diverse preferences. In any case, there is no real identification with the rights of others to freedom or equality because there is no passing beyond oneself to share the interests of descriptively *distinct* individuals.

The Rawlsian foundations for morality typify that way of thinking in which nothing is known directly and in which the individual stands alone, isolated from society, and finding nothing of real value. The 'Original Position' is perhaps no more than an ingenious device to overcome artificially the problems for morality which arise from our perceived isolation and resulting egoism, by demanding the formation of our moral principles in circumstances where we have surrendered our identity and merged with others. The fact remains that once we conceptually leave the 'Original Position', other people will once again appear as distinct and distant so that we remain fundamentally unpersuaded of the reality of their needs and of what is required for morality.

Modern liberalism demands no more than a counterfeit morality in which next to nothing is required and the little that is required is not active and positive (as in a genuine, full-blooded morality). It merely requires one to refrain from interference with the basically arbitrary choices of others, but no reason can be given why we should want to interfere with the choices of others, given that theirs, like ours, are arbitrary and interfere with no-one else's freedom.

As a result of this seductively undemanding morality, it has now become possible for people to *apparently* assert the truth of their beliefs without simultaneously asserting that they are true for anyone else, whether they believe it or not. The 'hard core' of belief has been removed when one's affirmation of it rests so naturally besides the denial of its truth for other people. There is no serious commitment in holding such beliefs and they can have no real power in guiding actions since one's arbitrary, subjective allegiance will not entail, or provoke, the striving against the negation of one's values. It is this unwillingness to strive against what contradicts one's will which Friedrich Nietzsche finds so repellent and life-negating. The ethical relativism of modern liberalism is arguably what he has in mind when he speaks so contemptuously of the insidious, 'last man' morality, and of the triumph of the 'mentality of the herd'.[14] It is *this* feature of alienated life which Nietzsche anticipates, and which he justly attacks because it represents a withdrawal from life, the needless resignation of the will, leading to indifference and despair.

Thus, we may interpret the 'fake' morality of modern liberalism as the Schopenhauerian side of the possible responses (discussed earlier) to the problem of alienation.

Notes

1.   The failure to act in accordance with a recommendation that one assents to can be put down to weakness of will. However, this does not explain why, in some cases, we seem to suffer from a weakness of will whilst, at other times, we demonstrate great *strength* of will. Difficulties in explaining why we sometimes fail to act on what we believe are due to the inadequate characterisation of a rationally compelling argument.

2. Kant, I., *Groundwork for the Metaphysics of Morals*, 'The Formula of Universal Law', §52.
3. Hare, RM., *Freedom and Reason*, Oxford University Press, 1965.
4. Rawls, J., *A Theory of Justice*, Oxford Clarendon Press, 1972.
5. Rawls is discussed further in section 3 of this chapter.
6. Plato, *The Republic*, Book 3, §406.
7. Aristotle, *Metaphysics*.
8. Husserl, E., *Crisis in European Sciences*, Northwestern University Press, Evanston, Illinois, 1970
9. Mill, JS., *A System of Logic*, Longmans Green & Co., London, 1879. Mill argues for a broader conception of logic in vol. 1, the introduction.
10. *ibid*, vol. 1, intro. §5.
11. Mitchell, B., *Law, Morality and Religion*, Oxford University Press, London, 1967.
12. Chapter 2. 'Of the Liberty of Thought and Discussion', J. S. Mill, *On Liberty*, Everyman's Library, London, new edition 1972.
13. Rawls, J., *op. cit*
14. Nietzsche, F., *Thus Spoke Zarathrustra*; 'Prologue', *The Gay Science*.

6 A Closer Look at Scientific Method

1. From Means/Ends to Programme/Prognosis: Myrdal's Solution to the Problem of Bias

In examining the problem of apparently irresolvable conflicts within the domain of value-neutral theory, I have focused attention on the role played by the interpretative framework in the formation of social (as well as natural) scientific theories. The 'interpretative framework' refers not only to items of supposed fact which the individual believes are beyond reasonable doubt (items which make up our common-sense, or background, knowledge) but also to judgments of value which determine:

(a) the selection of data,

(b) the relative significance we attach to data, both with respect to ourselves and to other data in our possession, and

(c) the way in which we employ data to give content to, and shape our action-guiding principles.

The value-judgments involved in interpretative procedures may be moral, political, intellectual or aesthetic; they may be consistent or inconsistent, concealed or explicit, but they will nevertheless play a decisive role in the integration of our experiences into a complex theoretical system which will enhance our knowledge and facilitate action. Where the social scientist unnecessarily handicaps him/herself is in denying the valuation-component enmeshed in theoretical knowledge itself. Thus, conflicts between theories which have their origins in differing interpretative frameworks will not be resolved where the opposing sides insist on the value-neutrality of their theories: neither side will address itself to the ultimate source of disagreement since to do so would carry the participants outside the scope of what they identify as properly 'scientific' debate.

A common response to accusations of concealed valuations which undermine the objectivity of competing theories (and hence diminish the chances of agreement) has been to strive all the harder to separate logically the valuations which infiltrate research, but which, on completion of the purely scientific aspect of an enquiry, may be 'tagged on' to produce informed decisions. Implicit within this approach is the means/end dichotomy discussed and assessed earlier (Chapter 1, section 2): it is assumed that the choice of ends, or objectives, is not capable of being settled objectively, by scientific method, but that the means leading to any posited end *can* be scientifically ascertained. Furthermore, it is assumed that values are not attached to the means themselves but that we are indifferent between alternative means except insofar as one or other has been established as most efficient in achieving a given end.

Underpinning the means/end model of scientific enquiry is what I have called a 'non-interactive' account of knowledge, an account which does not directly link the process of learning (including theory formation) with the desire to behave rationally. In an ultimate sense therefore, behaviour is seen as beyond rational justification – or at least beyond any justification which can be provided by an objective, value-free science – though in a less significant sense, the means/end dichotomist may concede that behaviour is rational insofar as the ends chosen are pursued, and attained, by following the most efficient means.

The maintenance of the means/end distinction depends upon a belief that we can gather together a body of facts (even gather and arrange them into a coherent theoretical *system*) in a wholly impartial manner, with the enquiring subject contributing nothing but its own detachment from the object of study. The object will, in turn, give up enough of itself to allow it to become genuinely known to the dispassionate observer. Once the task of impartial study is complete, scientists may then, if inclined, wonder precisely what to do with a store of objective knowledge acquired in so disinterested a fashion, but if they do, they will no longer be acting in their capacity as scientists – they will be looking at ends rather than studying the facts and causal relations which throw light on efficient means.

Although undoubtedly keen to stress the impartiality of the scientist, means/end dichotomists would not typically deny that value judgments have no part in 'getting science started'. They would concede to the critic that, at least at the outset, scientists have an idea of what sort of problems they (or their source of funding) wish to solve, and would agree that these problems will have been identified with reference to a framework of value-based objectives. Thus, for example, the natural scientist might look into alternative ways of generating energy for locomotion, given that locomotion is valued, and given that those who value it anticipate problems in connection with existing sources of fuel. Likewise, the social scientist will not randomly go out 'shopping' for facts; faced with social conditions which are negatively evaluated (e.g. unemployment) academic attention will be focused upon that sphere of reality wherein the causes of the perceived problem are to be found. It is commonly believed that, whilst our interests in certain aspects of the objective world are value motivated, our studies need produce no less an objective account of the world for that reason. Valuations may be prior to any form of scientific research in that they may motivate and direct it. They may also follow in the wake of research when, with the benefit of established fact, we are able to make informed judgments and perhaps even amend our choice of objectives. It is thought that no valuation, however, need *permeate* research, thereby violating its objectivity and compromising the scientific integrity of those concerned. The facts themselves, though coming under the light of directed awareness, are nonetheless known for what they really are, unshadowed by the light which falls upon them.

Means/end dichotomists no doubt feel that they can make this concession to those who argue that the scientist cannot be completely disinterested: *of course* the researcher needs a purpose, a value-based reason, or cause, for engaging in research activity. The dichotomist need only deny that the initial source of motivation does any more than to get the work started. The value-based motivation does not, or need not, affect the actual nature of the work undertaken. I now wish to suggest that this view only appears plausible if we misrepresent the nature of the judgments involved in research activity.

A theoretical system does not generate itself, nor can it be inspired by the disinterested observation of empirical phenomena. Rather, it is the case that scientists must set to work their faculties of judgment to achieve a theoretical, explanatory system constructed, not entirely *out of*, but at least *on the basis of* available data. The means-end dichotomist cannot deny that there is an ineliminable role for judgment in theory formation, a role which cannot be filled by simply consulting or discovering more facts about the world. Judgment may be sound or unsound, but without it science (if it could be anything) would surely be no more than a catalogue of observed phenomena, lacking even conjectural hypotheses about the relations *between* phenomena. As such, the body of scientific knowledge could have no significance for human action since it would tell us nothing about the effects of any action (or controlled change) on the environment. Action-guiding knowledge requires a theory, or explanatory system, however elementary this may be, and this, in turn, will require the exercise of our faculty of judgment. If the defenders of the means-end model of scientific enquiry wish to retain their value-neutrality, it would seem as if they must abandon the most crucial aspect of scientific activity, that which consists in theorising, explaining and predicting. If, on the other hand, they wish the findings of scientists to retain their significance for human action (and for social science to retain that element which renders it capable of informing social policy) then they must abandon all claims to value-neutrality, together with the means/end distinction designed to secure this.

One possibility remains for the upholders of the means/end dichotomy: they may indicate (as I already have) that there are many forms of judgment, or valuation, besides the moral judgment which is the specific type they seek to expunge from objective, scientific theory. Scientists can, and must, strive to eliminate sources of bias creeping in from their subjective moral valuations but need not (arguably *cannot*) avoid making what we may call 'intellectual' valuations which enable them to construct coherent theoretical systems. In contrast to moral or even aesthetic judgments, which indicate what does or does not accord with taste, the intellectual judgment, if it is sound, is meant to indicate what accords with truth. In this way, the possibility and relevance of theory to action is secured, whilst the objectivity of scientific theories relying on intellectual judgment remains unimpaired.

It appears the means-end dichotomist must now find grounds for distinguishing between two types of judgment: one subjective, being essentially an expression of taste, the other objective, expressing an intuitive grasp of some external state of affairs. May we not also challenge the basis of this new dichotomy?

In both cases, that of moral and of intellectual valuation, the facts fall short of settling the issues decisively and this is precisely why a judgment, or choice, is required. In what sense then can one particular form of judgment be said to accord with objective truth (that is, with a real, external state of affairs) and how could we prove that it did so? When the power of discerning truth from falsity passes away from the object to the perceiving subject of phenomena then the judgment as to what is true can no longer be called 'objective'. It is partially so in that the judgment is based on facts, but it has now acquired a subjective component which is reflected in the subject's choice. It would be quite arbitrary to say that the choice when it concerns the settlement of an intellectual question is inherently more successful in approximating to truth (and therefore more 'objective') than judgments on moral issues which are clearly not settled by facts alone. The feeling that the former of the two choices is somehow more objective is simply an expression of prejudice resulting from the deeply-entrenched fact/value distinction. Scientific theories are thought to be concerned exclusively with the realm of fact, forgetting that they too require valuations, whilst ethical assertions are traditionally placed firmly on the subjective 'value' side of the great fact/value divide. The similarities in the reasoning processes involved in both are often overlooked, either out of a misplaced desire to elevate science (expressing an epistemological need for secure foundations), or else to demote ethics and hence weaken the claims of objective morality which are felt to be inconsistent with the modern liberal foundations for tolerance. Since judgment steps in where the facts fall short, so to speak, it is in the final analysis the personality of the enquiring subject which determines in which theory's favour the decision falls, and as such, many considerations may be brought to bear on a question which are not necessarily illegitimate. If we understand by 'bias' not the mere fact that valuations play a role in discerning truth (which, I have argued, they must), but the circumstance wherein factual data is *pre*-judged so that the facts are not allowed to settle the question *so far as they can* (judgment intervening prematurely) then provided we are on our guard against bias, subjective factors may have a legitimate role to play in knowledge acquisition.[1] This is not to say that *any* subjective factors are admissible, that there is no such thing as *poor* judgment, even if we cannot demonstrate its poverty by proving that it is also the cause of bias. Both poor and sound judgment are often incapable of assessment by reference to readily observable facts, which is why sound judgment cannot be easily taught.

A sound judgment may be interpreted within the context of this work as one which stems from the fuller involvement of the subject with the objective world, so that what I subjectively apprehend is also what is objectively true. To recapitulate (Chapter 5, section 1), *my* world becomes *the* world in the act of knowing it as it is. Poor judgment, on the other hand, may be taken to be another form of alienated choice, stemming from the lack of involvement of the individual with the world in which s/he lives and seeks to understand.

Poor judgment is a condemnation as applicable in the realm of science as it is in art and morals.

To return now to the implications of this account of judgment in theory-formation for the 'means-end' defence of scientific neutrality. Given that there *is* a role for subjective valuation in the formulation of explanatory systems, it can no longer be maintained by the dichotomist that scientific theories are 'neutral' in the required sense. Science can have a value-independent subject matter but not, it seems, a value-free method of coming to understand that subject matter for what it really is. It is our *coming to know* the facts, if not the facts themselves, which cannot be judgment or value free.

The crux of the problem for the means-end dichotomist is that s/he can distinguish neither one class of judgments from another as being more 'objective', nor can s/he isolate so-called 'means' from 'ends' given that the identification of what is to count as means involves a valuation. This is so particularly in the social sciences, where the material under investigation is human behaviour and society, and moral values attach themselves to means as if they were simultaneously also ends in themselves. Thus, contrary to the dichotomist's protestations, valuations will not only precede and (psychologically but not *logically*) result from scientific research, but will permeate it also, and in such a way as to make theoretical science a possibility.

Recognising the profound difficulties for the means-end model in upholding traditional conceptions of scientific rationality, an alternative, more sophisticated model has been suggested by Gunnar Myrdal, a prominent social scientist who, coming from economics, is acutely aware of the complex intermingling of values with what purports to be objective economic theory. In his *Political Element in the Development of Economic Theory*[2], Myrdal identifies what he sees as 'the last remnants of metaphysics' entering economics more or less surreptitiously via the philosophies of natural law and utilitarianism. Whilst acknowledging the great and beneficial influence these philosophies have had over the development of economics, he argues that for the attainment of a body of positive economic theory, these metaphysical remnants must now be exposed and expelled. He is not content, however, with merely identifying the barriers to further progress but strives in the *Political Element* and subsequent work to provide a method by which concealed valuations may be brought to the surface, thus neutralising their corrupting effect on what would otherwise be objective theoretical knowledge.

At the outset, Myrdal rejects what he calls 'naïve empiricism', the view that if we observe reality without any pre-conceptions, the facts will miraculously arrange themselves into an explanatory system. This is the view which underpins the more crude means-end model of value-free science. Instead, he acknowledges that facts are only discovered and organised into conceptual systems according to the questions asked, and that the questions themselves are an expression of our interests, and hence of what *sort* of answers we require. Valuations, therefore, are taken to be involved even at the stage of observing facts and of embarking upon theoretical analysis.

The problem as Myrdal sees it is to show how subjective valuations are intertwined with the body of objective economic theory such that this knowledge retains its relevance and claim to be authoritative in the process of policy formation, whilst, at the same time, deriving and retaining its authority from its objectivity. To achieve this status, valuations motivating and structuring research must be made explicit, not merely at the outset, but throughout the work 'kept conscious and in the focus of attention'[3] so as to protect us from the subtle corruption of bias.

In the prologue to *Asian Drama*[4] Myrdal advocates that social scientists take a more active interest in the sociology of knowledge for the purpose of sharpening their awareness of cultural, political, environmental and psychological influences operating on their work. Such a study would better equip social scientists in their efforts to distinguish the value premises implicit within their field of study. Similar investigations have been carried out with regard to literature and art but rarely, to Myrdal's knowledge, with regard to the impetus for creative activity in social science where, arguably (given its claim to authority based on alleged value-neutrality), it is most urgently required. In addition to a sharpened awareness of the valuations influencing their own activity, individuals who are engaged in social research must acquire accurate information about the valuations directing other people's activity, and how these valuations in turn are affected by changing beliefs about the factual state of affairs. Prevailing patterns of valuation, and in particular, those belonging to the significant or influential social groups[5], should not merely be guessed at, or intuited, by the social scientist but uncovered as far as is possible by more extensive opinion research than is currently undertaken. Thus, the value premises chosen will not be selected arbitrarily; they will be empirically ascertained and made explicit rather than assumed as self-evident *a priori*, universally valid, and hence uncritically worked into the analysis. Since, as a matter of fact, many conflicting sets of valuations are held by groups in society, Myrdal further recommends that a number of sets of co-existing valuations be introduced as premises in order that the analysis will not appear one-sided when it comes to drawing out practical implications. In this way, the social scientist may be freed from the distortions of a source of bias which so often goes undetected and which, by muddying the waters of social scientific debate, makes agreement upon a core of objective, theoretical knowledge difficult to achieve. Selecting the value premises consciously, and then exposing them to the light for public scrutiny, clarifies the issues under debate, and not only the positive, scientific aspects of the debate but the moral values which are at issue also.

For Myrdal, it is important to recognise that since social research is to be carried out in relation to a set, or number of sets, of currently-held values, these valuations, forming the premises for a theory, will also circumscribe the limits of its relevance and objectivity. Social science must therefore abandon its claim to universality and hence to being a 'pure' science, but in so doing, the way will be cleared for a modest, more feasible, attempt at objectivity in understanding the changing principles of socio-economic interaction.

Evidently, Myrdal has gone much of the way towards tackling the very real and distinct problems faced by social scientists in contrast to their counter-parts in the natural sciences – the human subject matter of social science is capable of value motivated action, whereas events observed in the physical world are unmotivated, but, it is assumed, not uncaused. Hence, in the latter case, the search for unchanging, eternal laws is thought appropriate whilst in the former sphere, that which is of interest to the social scientist, the (descriptive) laws of interaction are prone to transformation as the valuations motivating human action are themselves transformed by people's changing perceptions of reality. Clearly, then, social scientific theory cannot be 'objective' in the same sense as the natural scientist speaks of objectivity, in the sense which implies the hope that a true theory is eternally true. Rather, Myrdal suggests, it may be objective in this restricted sense, that *given* the prevailing set of valuations, and *given* that these are, through empirical research, correctly apprehended and are both relevant and causally significant, *then* a theory so derived will present the most realistic picture of the outcome of various changes (policy decisions) introduced into the system. The main advantage of this over the means-end model of social scientific objectivity is that it allows for the complex interplay of ends, based on subjective valuations, with the means selected for attaining these ends. Myrdal rightly points out that valuations are unavoidably attached to the choice of means employed in social processes because the means, more often than not, will involve human labour and other forms of social or personal sacrifice. In addition, no objective will be desired by society come what may, regardless of the undesired side-effects produced in the effort to attain it. There will be a constant give-and-take between a wide variety of goals and an infinitely wide variety of methods for achieving those goals.

Feeling that the means/end dichotomy has to be rejected, Myrdal develops a more sophisticated analysis of the relation between value-based objectives and the factual state of affairs as studied by social scientists. The distinction between 'programme' and 'prognosis' is introduced which in some ways is similar to the distinction between ends and means but is intended to capture the idea of the responsiveness of the content of a programme and of the structuring of a prognosis, one to the other. The concept of a 'programme' includes the more complex set of ends, acceptable means and procedures which are generated by valuations; the programme will to a certain extent direct and shape research but it will also be affected by the prognosis insofar as values supervene on our perception of the facts. In other words, what we desire may alter when we become more fully informed of the implications of pursuing our initial objective.

A 'prognosis', of course, is the attempt at forecasting the turn of events based on past experience and observations. It requires the assumption that empirical study and sound judgment will yield up generalisations concerning the connections between actions and events which will prove useful in guiding future activity. However, those responsible for prognoses must be prepared to amend a prognosis in accordance with social transformation, reflected in the changing content of the programme. Thus, if the results of analysis indicate that in prevailing circumstances, a certain course of action conflicts with desired ends, and this information becomes generally known, it is possible that everyone will cease to behave in this way and thereby substantially alter the society under investigation. This, in turn, may partially invalidate the original prognosis.

It is worth emphasising at this point that, for Myrdal, there is no logical link between alterations in our factual beliefs and the valuations which

103

generate the content of a programme. If this were not so then programmes based on false beliefs could be logically reconstructed as further empirical study rectified the short-comings in our factual knowledge. The relations of values to facts is considered to be fundamentally arbitrary so that any changes in the pattern of valuation are seen as a matter for sociology, for sustained opinion research, rather than for logical derivation.

Myrdal's insistence on this point draws attention to what appears as a puzzling inconsistency based on differing interpretations of the demands of logic. When he asserts the need for opinion research on the grounds that values cannot be logically derived from facts, Myrdal presumably means that values cannot be strictly *deduced* from factual knowledge, which of course is perfectly true. As I said earlier, moral valuations involve a choice, which is required because the facts alone fall short of settling questions decisively. But is this not also the case with other forms of judgment, such as those involved in applying concepts or in selecting the appropriate tools for analysis? The substance of such judgments cannot be *deduced from* anything in a logically unassailable manner but rather must be inferred by the enquiring subject in such a way as invokes either censure or praise for poor or sound judgment. The fuller conception of logic discussed earlier (Chapter 5, section 2) is one intended to cover not merely deductive reasoning, but the art of inference whose mastery is the essence of sound judgment. When logic is thus conceived, it is not so clear that Myrdal is justified in distinguishing only the value premises (based on opinion research) as the subjective component setting the limits on an otherwise objective body of knowledge, for within the prognosis are incorporated forms of judgment which also will not pass the test of deductive logic. If Myrdal admits these forms of judgment (let us call them intellectual judgments) into the development of prognoses then he is accepting a broader conception of logic, one which he simultaneously wishes to deny when identifying the content of programmes as inherently subjective (that is, beyond reason and logic.) Thus Myrdal alternates between the fuller and the more restricted sense of what can be rationally justified.

In common with the means/end dichotomist, Myrdal shares a deeply rooted philosophical commitment to the fact/value divide, a division which runs parallel to the distinctions between objectivity and subjectivity, the rational and the non-rational. It is this commitment which makes it impossible for Myrdal to tackle the problems he sets out to solve: to show how economic theory can have implications for government policy and yet retain the objectivity from which its claim to authority is derived. When one's conception of objectivity is aligned with the rejection of the idea of rationally determined values then the problem Myrdal poses becomes irresolvable. This is because the nature of the objectivity demanded from social science excludes the possibility of science becoming theoretical, and hence of its having implications for rational action. Whilst remaining a moral relativist, Myrdal goes much of the way toward providing an interactionist account of knowledge: the programme/prognosis model reflects the interplay between facts and values and describes the steady acquisition of knowledge and the process of behavioural adjustment in a way that rings true. However, in this particular interactionist-sounding account of research, the subject's interaction is shown to fall short of violating the traditional conception of scientific objectivity. It goes far enough to apparently bestow relevance for action on the knowledge obtained but is then arbitrarily cut short so that the subjective component in acquiring knowledge is denied from the point where the core of objective, scientific theory is said to begin. Supporting this arbitrary cut-off point is the belief in a two-tier system of judgments, whereby those relating to moral issues are subjective, reflecting personal or cultural tastes, whilst those relating to intellectual questions are made honorary members of objectively true theoretical systems. This two-tier system is perpetuated by the deeply entrenched fact/value division referred to earlier and by the fear that if the subjective component smuggled in by intellectual judgment is acknowledged as such, then the whole corpus of scientific theory will collapse into the same abyss of non-rationality where morals are thought to reside. The possibility of rejecting ethical relativism and of revising the traditional Cartesian conception of objectivity, is not seriously considered.

The question now arises: *why* is this possibility not considered? What is there in the traditional conception of objectivity (that is, one which demands observations without the perspective of an observer) which is held so dear by those who uphold it? A clue to the answer may be found, among other places, in Myrdal's own pre-occupations, expressed in the *Political Element* and elsewhere, with the need to secure a basis for the social scientists' claim to authority, or rather for their work to be authoritative, in shaping social policy. The achievement of objectivity, when this is alleged to consist in giving an account of reality from the perspective of *no particular individual*, provides some justification for proffering the supposedly impartial account on behalf of *all* individuals and for using it as the basis for decisions taken on everyone's behalf. Thus 'objectivity' in this sense means 'valid for all', whether or not what is claimed as objectively true coincides with what is subjectively apprehended by particular individuals. In contrast to this is another conception of objectivity where what is claimed to be known is known only by virtue of a particular individual's imaginative involvement with the world, making the world his or her own. The latter does not sanction action on the individual's behalf; what is known is authoritative because the individual, by sound judgment and careful observation, has grounds in personal experience for maintaining its truth and for allowing such knowledge to guide activity. On a fully interactionist account of knowledge, the only objectivity required (and, I have argued, the only concept of objectivity which is intelligible) is that which makes knowledge authoritative for the individual. Beyond that (short of rationally persuading others of its truth), knowledge could have no authority in shaping the lives of others and therefore is not of the kind which, if possessed by an individual social scientist, bestows any special claim to influence in government policy.

In the prologue to *Asian Drama*, Myrdal himself speaks of the close links between economics and government, not of course with any disapproval but, on the contrary, with a certain pride and a desire to make economics worthy of this 'special relation' by purging it of bias. Thus he observes:

> Economists dominate fact gathering and planning in every country... For more than 200 years [they] have advised both those in power and those in opposition; some of us have been members of parliaments and governments...By comparison, the other social sciences have been 'poor relations'. [6]

There can be little doubt that Myrdal requires the 'perspectiveless' view of objectivity to sustain this position but he fails to notice that in adopting this particular conception of objectivity he incorporates a pro-authoritarian/anti-individualist bias into any economic theory which is meant to be objective in this sense. It is not sufficient for him to say that a pro-authoritarian valuation may be included in the explicit value premises, selected on the grounds of its being almost unanimously upheld in a society where people really do prefer to leave things to those in power. This will not do, because the valuation sneaks in to shape Myrdal's actual methodology. It is *this* valuation which forms a crucial part of Myrdal's own interpretative framework and which leads him to tackle the problem of bias, and the shortcomings of the means-end model of value-neutrality, in the way that he does – that is, in a way which is intended to preserve a core of authoritative (because non-perspectival) objective theory.

In *Objectivity in Social Research* [7] Myrdal himself acknowledges the links between the methodology of social science, as he conceives it, and a particular form of authority-based organisation, namely, modern liberal democracy. In a totalitarian society, it is at least arguably the case that a means-end model would suffice to depict a social science which is instrumental to the ruling élite: the ends would be fixed by it alone, leaving conflicting valuations and the undesirable side-effects of its choice of means to be borne in silence by the powerless citizens. In a liberal democracy, however, social transformation stems from a much broader base of individual valuations, taking into account, as it does, the wishes of more people. This is why the programme/prognosis account of social scientific procedure is more appropriate and why it is thought more likely to yield a realistic picture of social and economic inter-relations.

It is one of the persuasive strengths of Myrdal's more sophisticated methodology, of his account of how knowledge of facts shape, and are shaped, by valuations in the on-going process of social adjustment, that it allows social theory to 'move with the times'; it allows for changing valuations and yet – because prognoses retain a supposedly objective core – it allows social research to form the basis for government decisions taken on society's behalf. It is a conception of social science which accords perfectly with modern liberal democracy, accommodating a multitude of diverse, conflicting, and often shifting valuations whilst lending credibility to government decisions in matters which the vast bulk of the populace are too indifferent to play their part.

2. Background Knowledge and the Rational Selection of Theories

In the previous section, I argued that for scientific theorising to be possible, and for it to be relevant to human life, the exercise of our faculty of judgment is required to assimilate data into potentially action-guiding explanatory systems. Raw observational data, prior to being incorporated into a theoretical structure, must, of necessity, be insignificant as regards action because purposive action only follows upon the awareness of the relations of things to each other and to the agent him/herself. It is the theoretical structure (our *explanatory system*) which fills in these relations, which adds to the observed data the unobservable and often causally complex relations between things we perceive. The theoretical structure consists, therefore, of what is observed and of what is not directly observable; it results from our filling in the latter according to the dictates of judgment and will be extended as the existing structure provides an indication of the causal significance we might attach to newly acquired data. Stated otherwise, possible additions to the theoretical structure may be assessed, among other things, by their consistency with prevailing theories. The accepted body of theoretical knowledge will impose certain limits upon how we may incorporate new data into the structure which has been built out of, and around, the old. The structure may even impose such limitations as to rule out the very possibility of our making particular observations, of our recognising them as either new or significant in some respect. It is for this reason that Feyerabend[8] has encouraged the proliferation of theories as being more conducive to, and consistent with, a healthy empiricism. An alternative theoretical structure may prove capable of unearthing observational data to which followers of another theoretical school have been blind. This implies that science can proceed counter-inductively (that is, from theory to fact) as well as inductively (from fact to theory).

In what follows it will be my purpose to lend support to much of what Feyerabend has said regarding the development of science and, following on from the last section which maintains that even our knowledge of facts depends on making the right (value-conditioned) judgments, to argue that progress in science is from theory to theory, where each theory embodies the judgments we have been disposed to make. In opposition to Feyerabend, however, I wish to argue that we *are* nevertheless able to make a rational choice between alternative theoretical systems.

The advocacy of theory proliferation need not imply adherence to any form of relativism and, in fact, can be made consistent with a belief in the possibility of objective knowledge. For fallibilists such as Mill and Popper the freedom to explore, to increase theoretical diversity, is the pre-requisite for a healthy academic environment and ultimately conducive to arriving at *objective* truths. What still remains unclear is how a belief in there being an objectively better, and therefore rationally preferable, theory is consistent with the view propounded here that science proceeds from theory to theory, *and*, moreover, that each theory inescapably contains a valuation component which enters the process of theory formation when we bring our judgment to bear upon the sensory data.

It will be apparent by this stage that I do not regard the presence of a valuation component in the construction of theoretical systems as necessarily detrimental to their objectivity. Rather I have tried to argue that the conception of objectivity which demands that the social and natural sciences be value-neutral, that they provide a picture of the social or natural world as if known from no-one's perspective, is fundamentally incoherent. It stems from a non-interactionist account of how knowledge is attained so that the enquiring individual is always set apart from the world s/he seeks to know, so that our ability to know anything is undermined and ultimately usurped by experts who claim what *they* know (because objective in the non-perspectival sense) as valid for all.

The belief that a theory of any kind, and however elementary, can be purged of valuations involves some form of self-deception, since it requires the refusal to recognise one's intellectual judgment as a subjective component (among others) in the construction of explanatory systems. Certainly, the necessary presence of a subjective component indicates that the traditional demands of objectivity cannot be satisfied, but we need not yet abandon hope of knowing the world *as it really is*. What we must abandon is the possibility of knowing the world as no-one can know it, that is, from outside of one's own perspective. A conception of objectivity which grants the legitimacy of the individual's perspective and which also grants that we may, by sound judgment, make the real world identical with our own, subjectively apprehended world is sufficient to endow knowledge with the authority to command and direct (what is thereby *rational*) activity. An individual may know the world more or less fully depending on the openness and extent of his/her involvement with reality, an involvement which will enhance one's judgment the deeper it goes, but it will not be for any external arbiter and least of all for the mythical, 'neutral' or impartial arbiter speaking in the name of an impossibly objective science, to decide whether or not the individual has 'got it right'.

In stressing the importance of the individual and of the oppressive nature of a conception of objectivity which allows the scientist to masquerade as an impartial judge, in urging the toleration of diversity and the fuller use of our inventive faculties, I find myself again echoing the ideas of Paul Feyerabend.[9] However, unlike Feyerabend, I can find little sense in valuing personal liberty *above* truth, or in valuing diversity for its own sake. Liberty is the pre-requisite for individuals to act upon the world in such a way as to come to know *for themselves* what the world really is. It would appear to us as less valuable if our liberty meant we were free to make choices which could only ever be arbitrary, for if we believe that there are no objective standards for guiding action we very soon prefer to surrender our liberty and cling, instead, to any external authority which will give to our actions a purpose.

Feyerabend appears to hold an existentialist conception of freedom, a conception which is so absolute it fails to recognise even the constraint of rationality. Such freedom has been perceived by existentialists themselves as a burden rather than as an asset, as a somewhat disconcerting circumstance which imposes huge responsibilities upon the agent. Finding no objective values in the world by which to orientate activity, it falls upon free agents to create their own values and standards for rationality and it is only in so doing that they overcome the experience of disorientation.

Nietzsche has often been associated with existentialism insofar as he too perceives that humanity might be free[10] (in this absolute sense) if we could but recognise that the values, or truths, we have taken to be objective are nothing but our own invention. I have observed elsewhere (Chapter 3, section 2) that Nietzsche's response to the vision of an indifferent, valueless world is needlessly aggressive, that it springs from an alienating epistemology that sets the knowing individual at a distance from the world he seeks to know. Feyerabend shares his epistemology in this respect, which is why his prescription for 'epistemological anarchism' and political relativism will cure neither science nor society even in the short term – quite simply, greater freedom will not be welcomed when it is simultaneously denied that there can be any objectively meaningful purpose in the pursuit of which this freedom can be spent. When the perceiving individual is set apart from what is perceived then it is of no surprise that we can find no values really 'out there', that the world, as Nietzsche himself believes, appears to us as indifferent and valueless. This is because it is undeniably true that a world from which we have been removed can have no value, all on its own, *for us*. It is only when we are put back into the picture – and we really *are* in the picture, being as much a part of reality as the things we perceive – that the world acquires objective value in its relation to us.

To deny that there are theories which it is rational to prefer, or that there are things which are to be universally valued, is to deny, firstly, that we are a part of the reality we hope to understand, and secondly, that the relation in which we stand to the world is at least partially, if not largely, determined independently of our choice. (For example, the presence of other people with interests which can be violated is a feature of reality whether I will it or not.) Thus we cannot arbitrarily choose to establish ourselves in any relation whatsoever to the world, such as that of self-appointed tyrant towards one's imagined subjects and kingdom, although within the limits of the context we inherit we *are* free to choose objectives which will be more or less rational depending on our awareness of the context in which our activities take place. This context, which includes the network of causal relations, and other forms of interdependency, in our natural and cultural environment is what makes certain choices objectively right or wrong and accordingly, certain kinds of behaviour either rational or irrational.

The mistaken belief that commitment to a value system results from an arbitrary decision is grounded in the idea that our relation to the world is entirely self-chosen rather than, for the most part, inherited. Obviously to a certain extent we may establish ourselves in a relation to society which involves generating new moral commitments, as when we have children or freely promise to do something for somebody, but our obligation to avoid harming people or destroying the environment is not of this kind. Our obligation to avoid doing harm persists regardless of our choosing to acknowledge it for ourselves since it is not brought into existence (like the obligation associated with a promise) by the contingent act of our choosing to value another person's well-being. The value of another's well-being is quite independent of our opinions on the matter in a way that our making a promise is not, and if we failed to grasp its importance, it would be because there is an aspect of reality (someone's suffering) which we have not fully apprehended. Under these circumstances, the actual world would not be identical with *our* world in every respect.

It may be worth emphasising that simply because there are some things which become objectively valuable (as when we make and keep a promise) this does not mean that there are not also things which have always been of value, which requires no decision to generate an obligation, and which no individual has the power to rob of its value by simply withholding his or her acknowledgment of this fact. There is a context for activity which we ourselves create (this will be some element of the social or man-made context) set within the broader (natural world) context, which is largely given, and to which, taken as a whole, we will respond with a greater or lesser degree of rationality. Individual liberty appears only at its true value alongside this recognition, that there *is* a more and a less rational way to conduct one's life, both in the choice of, and adherence to, self-made commitments and in the recognition of those commitments which the given context for our activity thrusts upon us.

To return again to Feyerabend's conception of a free society. When the choice between alternative world views, scientific theories, or systems of value is seen as ultimately non-rational then however highly we hold freedom as an ideal, it will be esteemed far less in practice. Just as the tamed bird will not fly out of the open cage when he has been robbed of all experience of anywhere he might have wanted to go, so individuals, believing they have absolute freedom and the chance to exercise it, will see no good reason not to renounce that freedom and cling instead to whatever is familiar, for if there is one thing which is valued above freedom it is that our lives may have a meaning or purpose that is non-arbitrary. When this possibility is denied, liberty itself loses much of its appeal since it is prized, at least in part, as the pre-requisite for achieving objectively worthwhile ends.

So far, I have spoken only of how the objectivity of our theoretical (and hence practically relevant) knowledge is possible given a conception of objectivity which recognises the contribution of the enquiring subject. When the appropriate framework for interpretation is brought to bear upon all available data, we may say that the enquiring subject has sound faculties for judgment, that he or she is free from bias and ideological distortion (about which more will be said in Chapter 7). The knowledge so acquired may then be deemed 'objective' in a non-Cartesian sense, meaning that it is not, and cannot, be purged of its subjective component because the latter is what is required for the knowledge to have practical import. The so-called subjective component enters into knowledge most clearly whenever what is known is intended *as an explanation*, since this involves an attempt to grasp causal connections which are not directly observed but which must be inferred from the circumstances surrounding the event. Where the subjective component is less apparent, and even more frequently denied, is in the description of the observational data which forms the raw material for explanatory systems. This data, at least, is taken to be incontrovertibly value-neutral.

For many, the authority of science as a rigorously objective discipline which admits of the possibility of rationally determining the better of two or more theories (and hence of bringing us closer to the truth of how the world is) is thought to depend upon the neutrality of empirical data. Value-neutral observational data are taken to be the arbiter between competing theories. They form the testing ground for new hypotheses and as such qualify as the secure foundation for building the edifice of scientific enquiry. Raw empirical data, by virtue of their independence of the value system of the observer, are taken to be the touch-point with reality, the point at which good scientific theory intersects with the world it is meant to describe. Theories are therefore testable at every point where their implications connect with what can be empirically confirmed or denied.

It has been thought to follow from any attack upon the value-neutrality of raw data, that scientists must relinquish their claim to having a rational basis for their allegiance to one theory rather than another. When our understanding of the objective world is seen to be necessarily tinged with the subjectivity of the observer, both at the level of gathering the data to be explained and of incorporating them into the wider theoretical framework, then it becomes tempting to seek an explanation of the choice between competing theories and of the dominance of one scientific paradigm over another, in terms of the social and psychological pressures brought to bear upon the scientific community. In other words, once the secure foundation provided by supposedly value-neutral data is shown to have crumbled then the emergence and acceptance of scientific theories becomes part of the non-rational process of human development. The idea that science *reaches out*, bringing us closer to the external world, is then lost and replaced instead with the disturbing notion that our knowledge simply reflects something more about ourselves. Once again we begin to feel trapped within our own limiting perspective.

To escape this conclusion, we need not, however, return to the defence of the possibility of scientific neutrality or to the working out of a still more sophisticated methodology to refine the facts from the so subtly value-contaminated theories and data. Perhaps we do not need the 'uncontaminated' data after all but can find elsewhere a basis for the rational preference of one theory (or scientific tradition) over another. This is my contention and I would suggest as the candidate our common-sense background knowledge. It is Popper's recommendation too, though he only intends scientists to refer to those areas of background knowledge which could be deemed free from subjective valuations. But I disagree with him that there are such areas – much of the knowledge we take for granted *appears* to us as valuation free, only because it is so widely accepted and largely unchallenged that it rarely gives rise to value-centred disputes.

It is the essential insight of Popper's evolutionary approach that no body of knowledge is ever built up from scratch, as if each generation of scientists must begin like ignorant Adams and Eves with nothing to inherit from their ancestors and everything still to learn. Knowledge acquisition is seen by Popper as an on-going process of extending what we already know, and sometimes even revising what we take as knowledge in order to better accommodate new data. A characteristic of this approach is that knowledge acquisition is taken to be an activity of the rational subject upon the world s/he seeks to know; knowledge is not seen as something to be simply poured into us via the five senses as if we were no more than passive receptors – or 'empty buckets' as Popper puts it[11] – but rather we are required to *do* something in order to acquire knowledge. We bring our background (and still more tentative) knowledge to bear upon new data and understand them through their relation to what we already know. As living beings we are always involved in some form of activity, there is always a need to respond to changing circumstances, and so we cannot face life 'empty-headed', as it were. We enter the scene with pre-conceptions, or, as I would prefer to put it, with our ready-made framework for interpreting experience, and in so doing we are better able to respond *even when our framework proves inadequate*. Frameworks can be tampered with, and in time they may even be thoroughly transformed, but they can never in one fell swoop be demolished before another interpretative structure has been formed to facilitate rational behaviour. The reason for this is that human activity will not go on hold. Life will not wait until we have re-built our interpretative structure from scratch, and to proceed as if it did, abandoning one perspective on the world before we had gained another, would not be thorough or particularly cautious but downright irrational. We would be left blind in the interim period, stumbling around and comprehending nothing. What is more, from such a position, we could not even hope to re-build an improved explanatory system. This is precisely why Popper believes science must assume common-sense realism for without it, science could not get started.

I have already said that I do not demand or expect background knowledge to be free from subjective valuations in order to be relevant for scientific enquiry; nor do I expect it to be acknowledged as an inflexible, totally unchanging part of our total body of knowledge, not prone to the amendments that affect more tentative areas of knowledge when recalcitrant evidence is amassed threatening their cogency and explanatory power. By 'background knowledge' I mean *not* the sum total of all our knowledge, together with the most tentatively espoused theories and facts that may be brought to bear when assimilating new data, but only the 'hard core' of knowledge which we feel most disinclined to abandon. Today, this will include, for instance, the belief that the earth is spherical and orbits the sun. This is an item of common knowledge which was once thought as ridiculous as it is now thought obvious. In fact, the historically prior idea that the earth is flat now seems to us almost absurdly counter-intuitive, which illustrates the point that although science must begin with our common-sense notions about the world, it *can* go beyond them, probing deeper, and ultimately transforming what we take to be common-sense. (In saying that science must assume naïve realism we are not asserting that it can never overturn our initial, naïve perceptions. Science results from our perceiving *more*, not from our perceiving in a new, non-direct, non-reality-encompassing way.)

What this means is that our judgment will naturally err on the side of demanding that new explanations be consistent with at least the hard core of our body of knowledge, rather than the converse, that we should immediately consider tampering with our background knowledge in order to make the new hypotheses fit. On the face of it, this sounds like a bias *against* progress and *towards* maintaining the status quo – and of course in a sense it is – but it is an entirely rational bias (if we can still call an attitude that is rational a 'bias') for the reason stated above, namely, that we cannot abandon one explanatory framework for another that is as yet too insubstantial to support rational activity.

Background knowledge is what we must take for the time being as the bedrock of all further enquiry and as such, it can also, in one sense, serve as the neutral arbiter between two theories, or traditions, which are at odds with one another. It is important to emphasise that 'neutral' here does not mean 'value-free' because background knowledge, like all our knowledge, will have a subjective component (being in part the product of our sense of judgment); rather the term 'neutral' will now be taken to mean 'impartial', that is to say, independent of the theories being compared and tested.

Sometimes, in a period of scientific revolution, the new theory or approach being urged will consist of denying some portion of the theories which have acquired the status of background knowledge – the belief that the earth is flat is one such item of knowledge challenged and overthrown by the Copernican revolution. But losing some part of what was once background knowledge does not mean that we are unable to assess rationally which is the better of the two world-views. Nor does it mean that Copernicus must be seen as necessarily irrational because he rejected a crucial part of the old interpretative structure. Enough of the original structure will always remain to guide the scientist and the public through the period of transition, indeed to provide the basis from which the enquiring mind can reach out and grasp the newly emerging ideas. Thus, 'Old World' remnants may, for a time, form part of the common ground between old and new world views and serve as the basis from which to launch further enquiries even if, ultimately, these enquiries lead to their expulsion from the corpus of both 'background' and more tentative knowledge.

We can only rationally eliminate old theories on the basis of what we know/are coming to know. Thus, in the early days, the ascending tradition may provide insufficient grounds for the elimination of certain theories. The measure of the success of any theory is how it fits in with our common-sense background knowledge. As the latter changes (which it does in a period of revolutionary transition) then the 'neutral' basis for assessment is extended and eliminates further aspects of the old explanatory structure. Theoretical remnants from the demised structure may begin by losing their status at the hard core of the explanatory system, perhaps because they are no longer protected by those areas of background knowledge which have been cast out. This means they have become more vulnerable to recalcitrant evidence and, in time, they too may be eliminated.

It is of no surprise that students of the history of science have been able to detect both continuity and discontinuity over the revolutionary divide: our sense of the discontinuity is enhanced when we look at those areas of background knowledge which have been transformed, and when we appreciate from the point of view of the revolutionary participants the personal upheaval this must have involved, whilst our sense of continuity owes much to the fact that something remains of the old and lives on in the new to provide the basis for rational comparison and theory selection.

Koyré's[12] account of scientific development acknowledges this dual aspect to revolutionary change: new ideas are understood to emerge *in opposition* to the old, not spontaneously, but gradually, as part of a struggle against the limitations of the obsolete system of thought. The process takes time, and in retrospect, the transition appears discontinuous, whilst it remains true that, in the early stages, the new theories carry with them some remnants from the demised theoretical structure. For Koyré, it is these remnants which give the *spurious* appearance of continuity, concealing from the historian the true

116

nature of what has happened as a radical and genuinely revolutionary change. I cannot agree that the remnants of the declining school of thought give a *misleading* impression that there is some degree of continuity. If there is not really some continuity it would suggest that their presence in the early life of the new theoretical structure is purely coincidental, like dust in a cupboard which has not been properly swept. I am suggesting that these remnants play an important role in carrying the scientist through the revolutionary transition, enabling him/her to proceed rationally by delaying the demolition of crucial parts of the 'hard core' of background knowledge until an adequate reconstruction of our explanatory system has taken place. To do otherwise would be like pushing away our bridges before we have established a firm footing on the opposite shore. (If I may elaborate upon this metaphor, the opposite shore would represent the discontinuity between the opposing traditions, whilst the bridge signifies that in getting from one shore to the other, we have moved along a continuous path.)

To recapitulate, there are two ways in which the revolutionary transition achieves continuity:

1. from the point of view of the participant involved in the process of change, dismantling bit by bit, the central elements of the old interpretative structure and replacing them with new elements, and
2. from the perspective of the historian looking at both pre- and post-revolutionary science and being able to make a rational choice between the two on the basis of the background knowledge which is still common to both. If there were no items of shared background knowledge, a history of the development of human thought would not even be a possibility because the obsolete systems for explanation would simply be incomprehensible to us.

We will refer to something as a 'revolution' when, between two close, though distinct, points in human history, the make-up of our background knowledge is altered, thus radically re-shaping our framework for interpreting the world. Amendments to the more tentative areas of our knowledge, though also affecting the framework, do not constitute revolutionary change. It follows from what has been said that any debate about whether or not any change constitutes a revolution must depend on what is perceived to have been part of the 'hard core' of background knowledge at the time prior to an alleged revolution, and whether or not acceptance of the new theory implies a rejection of some part of this hard core. What is *not* required to establish that there has been a revolution (and the discontinuity in the process of change which this term implies) is to show that there has been *no* continuity, that everything that was once held to be true has been swept away to make way for an entirely new explanatory system, since to demand this of the revolutionary thinker is to demand, if not the impossible, at least the imprudent and the highly irrational. It is to demand that we build new explanatory systems not in opposition to the old, but from nothing, that is to say, not in response to anything which is *known*. Since all we can have of the world is what we can know of it, such a demand really is tantamount to wanting to bring forth something out of nothing. Knowledge does not come out of the state of pure and absolute ignorance, it develops in tentative stages from lesser or mistaken knowledge claims. This follows from the world being given directly in experience. We can either make the world as it is our own or fail to know what the world is in many of its aspects. We cannot in experience avoid it altogether.

In *Science, Revolution and Discontinuity*[13], John Krige raises the question of whether, and when, the revolutionary project is rational and argues that to sustain the view that the scientist's switch of allegiance *is* rational, it is not necessary to stress theoretical continuity at the expense of diminishing the revolutionary significance of the emerging ideas; that is to say, it is unnecessary ultimately to deny that there has been a revolution in order to maintain the belief that scientists in a period of rapid change have behaved rationally. The feeling that discontinuity in the development of science is incompatible with the exercise of reason is thought, by Krige, to stem from what he identifies as the liberal conception of rationality, a conception which demands that rational transition, whether it be in the scientific or political sphere of life, be evolutionary in nature rather than the result of an abrupt and violent rupture with the past. (The latter Krige identifies as belonging to the Marxist conception of rationality.) The conception of rationality which I have been promoting draws on elements of *both* the liberal and Marxist conceptions, uniting the diverse aspects of our intelligence. Thus, whilst I agree that there must be an element of continuity in thought, I also believe that partial continuity provides the secure basis from which the imagination can leap forth and grasp onto something entirely new. It is the leaps and bounds of the human imagination which give to the revolutionary transition its discontinuous aspect.

Even among philosophers of science who do not deny that there have been periods of discontinuous transition, such as Kuhn, the liberal conception of rationality remains influential. Thus, Kuhn in *The Structure of Scientific*

Revolution interprets the transference of allegiance from one paradigm to another as an event which cannot be rationally accounted for and instead he enlists non-rational sociological and psychological factors to explain a paradigmatic shift. For Kuhn, it is only during periods of 'normal science' that science can be said to develop rationally, in accordance with the central theories and methodological procedures that characterise the paradigm. The period of revolutionary transition – of non-rational *non-science* – is interpreted historically as swift and short-lived. It is essentially a 'gestalt switch' view of the moment of revolutionary change. Krige himself, because he rejects the liberal conception of rationality, is able, like Kuhn, to maintain that the history of science is subject to marked discontinuities in those periods described as 'revolutionary', but *unlike* Kuhn, he is not forced to misrepresent their nature as one that is neither historically protracted nor rational.

Adopting a 'Marxist' conception of revolutionary change, Krige presents the revolutionary process as a prolonged struggle on the part of the participants to disengage themselves from what they have come to perceive as the restrictions of the old theoretical framework. Even in the early stages, before an improved framework is fully elucidated, it remains rational to work towards the rejection and replacement of the old structure, given its increasingly apparent incapacity to solve certain problems, and also given an inspired glimpse of the new structure which, though underdeveloped, shows greater promise. Breaking with what is familiar and well-established becomes rationally justifiable once it has been seen to be overly restrictive. On the other hand, committing oneself to the unfamiliar is justified when it opens up possibilities for extending our understanding further than ever before. To be sure, there *is* an element of uncertainty in the revolutionary process but then there is in all our knowledge once we realise that we cannot have the epistemically secure foundations demanded by non-evolutionary, non-interactionist accounts of knowledge acquisition. Perhaps the only thing which is certain, prior to a revolutionary transition, is that the old theoretical structure *is* inadequate and therefore *must* be replaced as rapidly as human reason will allow – and this realisation would certainly account for the sense of urgency which is characteristic of the foremost revolutionary participants.

The important point which I think emerges from Krige's discussion of the revolutionary process is that it is only when we amend our conception of human rationality that we are able adequately to portray and comprehend the behaviour of scientists, and, what is most important, to accept their behaviour as at least conceivably open to rational explanation. Discontinuity in the development of thought implies an irrational 'leap' only when one's conception of rationality is unnecessarily limiting, when it is denied that acts of the imagination have a part to play in embracing new perspectives on reality.

For the most part, the liberal conception of rationality seems to cover the way in which science actually proceeds. Thus, scientists will extend the theoretical structure by working outwards from the central tenets which characterise the ruling paradigm and, following distinctive, shared procedures, will develop new theories and hypotheses. These latter will then have the character of having been deduced, or at least strongly implied if corroborated, because they have been attained by following the established rules of the scientists' game. During a revolution, however, the rules themselves are over-turned so that it becomes very difficult on the restrictive view of rationality to attribute to the scientist a rational motive. In contrast, the fuller conception of rationality (which has been discussed already) offers an interpretation of rational conduct which does *not* consist only in following rules, but in responding in a flexible manner to a world which can often surprise us. It involves the recognition that fixed procedures for explaining the world can only take us so far, and if clung to without compromise, can restrict the range of our possible responses.

If rationality requires our flexibility, our ability to adapt, amend, or overthrow our view of the world when it fails to 'work' for us, then rationality cannot be understood as the mere following of rules, however successful these have proven to be in the past. Breaking rules, abandoning old ways of thought for new, thinking imaginatively and creatively, must be options left open to the agent who does not wish to sacrifice rationality for the greater security of a 'tighter' argument. Once again, I am here echoing the ideas of Paul Feyerabend who has repeatedly argued[14] that rule-governed methodologies, though undoubtedly useful, should not be permitted to over-rule us. Rules are useful so long as they serve us, and use*less* once we allow ourselves to enter *their* service.

Characteristically, it is the more imaginative aspect of our rationality which is at work during periods of revolutionary transition, which is why such periods of intellectual progress are correctly described as 'discontinuous': to make leaps from one perspective, or interpretative framework, to another is the distinctive function of the imagination, of the reality-encompassing mental act. Of course, we may individually make such leaps during periods of 'normal' intellectual activity, though generally speaking these periods are more conspicuously characterised by those ways of thinking which are recognised, under the liberal conception, as definitive of rationality (that is, 'step by step' reasoning, with the associated characteristic of continuity). It is by comparison with these 'continuous' periods, where thought evolves rather than switches track, that revolution appears as the great upheaval that it is. Changes in our background knowledge alter our view of the world in a very fundamental way, so much so that to change our position requires argument which is emotionally persuasive as well as formally sound. (This point has been discussed in Chapter 5, section 1.) This is because common-sense background knowledge touches our lives at the very deepest level.

It is, however, unthinkable that all of our background knowledge could be removed in one go and to this extent the upheaval which signifies revolution is mitigated. No portion of what we claim as knowledge could be discarded unless it failed to cohere with some still more crucial part of our knowledge, that is, some more crucial part of our *background* knowledge, since this represents all that we *can* know about the world; there simply is no other standard for assessment, no mysterious 'world-in-itself' which is unperceived, unknown, yet somehow available to us to serve as the yardstick for measuring the truth of rival theories. Thus, even in periods of revolution, so long as some part of our hard core of knowledge remains, independent and untouched by the dispute, we have some basis for judging the success of the competing theories, and, ultimately, for making a rational choice between the alternate world views on offer.

Notes

1. The nature of 'bias' is to be discussed in Chapter 7.
2. Myrdal, G, *Political Element in the Development of Economic Theory*, Routledge & Kegan Paul, London, 1953.
3. Myrdal. G. *Objectivity in Social Research,* Wesleyan University Press, Middletown, Conn., 1969.
4. Myrdal, G, *Asian Drama*, Lane, London 1972.
5. Those with the power to translate their valuations into actions.
6. *ibid*, Prologue, § 8, p28.
7. Myrdal, G, *Objectivity in Social Research.*
8. Feyerabend discusses theory proliferation and the counter-inductive method in *Against Method*, Verso, London, 1988.

9. Feyerabend, PK, *Science in a Free Society*, New Left Books, London, 1978.
10. Nietzsche's views on human freedom are complex: true freedom is seen as 'the privilege of the strong' and it consists in noble self-mastery or inner constraint. The *ignoble* masses, on the other hand, present a rather more deterministic picture of human activity.
11. Popper, K., 'The Mistaken Common-sense Theory of Knowledge', 'Two Faces of Common-sense' from the collection *Objective Knowledge: an evolutionary approach*.
12. Koyré A., *Metaphysics and Measurement*, Chapman and Hall, London, 1968.
13. Krige, J., *Science, Revolution and Discontinuity*, Harvester, Brighton, 1980.
14. Feyerabend, P.K., *Against Method*.

7 Ideology and Bias

1. Defining the Terms

This chapter will be concerned with clarifying and exploring the concept of ideology as it has been employed in this work, and with contrasting my usage of the term with the ways in which it has been interpreted and employed by others. In addition, I should like to examine the relations between what I have now frequently referred to as 'ideological distortion', the concept of 'bias' (which, in Chapter 6, I suggest is distinct from ideologically distorted thinking) and the so far admittedly vague notion of 'poor judgment'.

I shall begin with the concept of bias. In Chapter 6, when considering Myrdal's response to the problem of bias permeating social scientific research, I argued that if 'bias' is taken to be the infiltration of the scientist's subjective valuations into the theoretical structure then bias simply cannot be avoided. Observational data alone will not arrange themselves into complex theoretical systems but do so only with the help of the scientist's faculties for discerning causal relations, for weighing up the impact of one set of events upon another, for assessing the significance (or relative insignificance) of given phenomena, or social groups, within the system as a whole. In other words, the construction of a theory which purports to explain the world requires the intervention of our faculties for judgment. The intellectual, moral or aesthetic judgments which feed into our explanatory structures constitute the unavoidable subjective component within any body of knowledge. To rid theory of this component would be (if it were even *possible*) to attain the perspectiveless perspective, the 'view from nowhere' over our world which could not even rightly be called a position of knowledge since it expressly excludes the presence of that which knows – the subject itself.

Thus I have rejected the definition of 'bias' as the infiltration of *any* subjective valuation whatsoever into the formation of theories, though I will maintain that some kinds of valuation (and not necessarily moral valuations) are nevertheless illegitimate. This ability to pick out and make the right valuations is what I mean by 'sound judgment', about which more will be said presently.

What then can be said of my notion of bias? In Chapter 6, section 1, I give an account of 'bias' as the enquiring subject 'pre-judging the issue so that the facts are not allowed to settle the question so far as they can'. This account, in the light of my later assertion (Chapter 6, section 2) that even facts are theory-dependent (and hence reflect *in any case* the input of an enquiring subject) must now be further clarified. Obviously I cannot be maintaining that *because of their value-neutrality* facts should be allowed to settle questions so far as they can, for the value-neutrality of our knowledge of facts is here denied. Why then should we be concerned to hold back the intervention of human judgment to settle issues where the empirical evidence falls short, when there is already a subjective component integrated into what we choose to call 'the facts'? The reason, I would suggest, is that of consistency. If by 'facts' we mean those (theory-laden) facts already acknowledged by an individual (perhaps as part of one's background knowledge) then to disregard these facts in certain contexts whilst they are so readily affirmed in others could be classed as bias. Thus, for example, there would be an inconsistency in the recognition of facts if a football referee who knows what sort of behaviour constitutes a foul, fails to describe it as such when the foul is perpetrated by his favoured team. Certain, possibly (though not necessarily) dubious valuations concerning his team (eg. their unfailing sportsmanship, their actually deserving to win) in this case will have prematurely intervened to settle the issue in defiance of what the referee would otherwise have taken to be the facts.

On this account therefore, bias is a sort of disregard for the facts, but not just *any* facts; it involves only the disregard of what is already taken as 'fact' by the biased person him/herself. Thus, if I do not quite know how to distinguish a foul from legitimate though aggressive play, or when a foul warrants a penalty decision, then, although I could be accused of ignorance or incompetence, I could not be accused of bias *even if* I make the same judgment regarding the course of play as the biased referee. My judgment results only from an inadequate knowledge of the game, rather than from the premature, or wholly inappropriate intrusion of certain of my valuations causing me to set aside the knowledge (of the rules of the game) which I already possess. Hence, we cannot say that it is the particular judgment which is biased but the individual who expresses the judgment. Bias may therefore be understood as a form of *personal* inconsistency resulting from the inconsistent application of one's *own* framework for interpreting experience. This means that, by comparison with the other forms of error which we shall be considering shortly, it is relatively easy to both identify and correct bias because what is needed to overcome the biased judgment is the recognition of certain data, given their usual interpretation and significance, by the person who *already knows* how to do this; that is to say, the relevant facts which will 'kick out' against the biased judgment are, given a person's existing explanatory structures, readily accessible to the biased individual him/herself, making the position of such judgments within a network of inter-connected beliefs an inherently unstable one.

In defining bias as a form of *personal* inconsistency I mean to distinguish significantly this kind of error from that which may be called 'ideological'. 'Ideological error' refers primarily to the choice of an overly-restrictive framework for interpretation (though we do apply the term to individual statements also – but more on this below) and as such it is not so much an error of inconsistency but of ignorance hiding behind a barricade of well-constructed, often highly *consistent*, claims.

The biased judgment errs (as our example above shows) only in relation to the judgment we would have been led to expect *given a particular individual's interpretative framework.* It follows that we may call a judgment 'biased' even if it is not believed to be incorrect. For example, supposing we know someone to be fanatically pro-monarchist. Then we learn that this same individual is expressing a surprising opinion: Prince E is a social parasite! Of course, we may agree entirely with this judgment, but we need not refrain from expressing the opinion that the pro-monarchist is actually *biased* against Prince E – perhaps for the reason that she finds him to be uncouth or unattractive or perhaps because, having once spent hours waiting in the rain, the prince had failed to single her out for a kindly word and a regal hand-shake. Similarly, it would not be thought an extraordinary usage of the term to describe a judge as 'biased' for finding a youth guilty of car theft even if we knew for certain that the youth *was* guilty. We might have grounds for thinking that the judge was swayed not by the available evidence as s/he claims to be but (e.g.) by the youth's skin colour, education or class background. Thus the accusation of bias is levelled after the assessment of how a particular judgment stands in relation to other judgments or declarations a person might make. To say a judgment is biased therefore is, primarily, to bring into question the consistency of the speaker and not the truth or falsity of what is being claimed. It is to say that the person making the judgment has insufficient grounds for *personally* believing it to be true, although there may or may not be grounds for *other* people to think that it is true.

In contrast, the second form of error which I shall be considering – error in choosing between explanatory frameworks, rather than in the application of any given framework – *is* of a type which can be uncovered by showing that the world-view in question is based on a misrepresentation of the world as it really is. However, this does not mean that such errors of judgment may be more straightforwardly identified than those mentioned above. On the contrary, ideological, reality-distorting frameworks have as their basis relatively stable, and often highly consistent sets of beliefs and it is for this reason that errors of judgment in the choice of framework cannot be so readily identified, nor so easily dispelled by what usually counts as an adequate rational argument.

Apart from the high degree of internal coherence which is characteristic of at least the more influential ideologies, there are other features which make the identification of ideologically-based errors more difficult. As was the case with bias, there is again no direct link between *any particular* judgment's being termed 'ideological' and its simply being factually incorrect. Consider, for example, the specific judgment 'Prince C is hard-working and conscientious. He often puts in a 16 hour working day'. Such a judgment, considered in isolation from other claims and evaluations a person might make, may well be a sound judgment of Prince C's disposition and personal qualities based on *known* facts about his working life. However, this judgment nevertheless gives support to the claim, which many people might make, that the speaker is in the grip of an ideology in which pro-monarchist beliefs play an important part. In other words, the specific judgment may be correct, based on established and accepted fact, and yet still be labelled 'ideological' because of the possibility that it is being used, for instance, to justify the existence of a monarchy. It would be the attempt at such a justification which would be regarded, by many, as misguided, though it is not necessarily the case that every specific judgment which lends support to the monarchy is, in itself, erroneous.

In practice we often find the term 'ideology' used in the derogatory sense, to denote our disavowal of the wider world-picture, and of the wider value-system, in which a particular, albeit true, judgment takes its place. So when we call a specific judgment 'ideological' we are, in some sense, saying that it is erroneous, clearly not in the obvious sense of its being unjustified by the known facts (which, in any case, it may not be) but in the sense that we feel it ought not to have been uttered in a particular context since it will be understood *within that context* as effectively saying much more about the speaker's beliefs than is given by the words alone; that is to say, we do *not* consider the statement in isolation from the speaker's other beliefs and evaluations and we therefore cannot help but feel that the speaker *is* in error if we, personally, have rejected the world-view to which he or she has revealed their commitment. Thus it is the world-view as a whole, and not necessarily the specific judgment (which has merely provided a clue to the wider belief system) which is thought to be inappropriate given the world as it really is.[1]

Confusion about the nature of the error being made when strictly true, though ideological, beliefs are expressed can lead to some perverse responses from those who have perceived the stated beliefs as 'ideological'. It may seem important to show the falsity of the other person's factual claim, to discredit the evidence which supports it, even though it is evidently true. Thus, for example, someone who perceives the statement 'Communism is associated with a generally low standard of living, and with some serious violations of human rights' as part of an ideological outburst may well feel under pressure to argue that, on the contrary, life in communist countries is good and that the stories of atrocities are all, apparently without exception, inventions of the Capitalist press. Thus one kind of error – that of working with a dogma, or ideology, which renders people incapable of seeing the flaws in their own society – is compounded by another, that of trying to dispel the afore-mentioned error by accusing its perpetrator of *every* type of error imaginable. It is as if those expounding what we take to be an ideology cannot be allowed to get *anything* right.

To clarify: it is not the relation of any specific judgment to the real world which makes it ideological but the fact of its being rooted in a restrictive explanatory framework which *does* misrepresent the way the world is. If we understand 'ideology' as referring to those explanatory structures which *limit* our understanding of reality, then, by implication, we are not claiming that possession of an ideology renders all of one's assertions false, but merely that one's claimed knowledge is necessarily incomplete and, to some extent, presents a distorted picture of the world to which the subject refers. Ideologies, in other words, are complex structures made up of both false *and* true statements. The truth of some claims may strengthen the position of those that are false. They are relatively secure structures insofar as they have a high degree of internal coherence and considerable explanatory power.

Although all the really compelling ideologies must embrace at least some beliefs which are true, it is also possible, and indeed it is often the case, that specific ideological judgments can occur in the absence of any evidence. In such cases, their basis for support is the wider belief structure, the ideology which they reflect. It is the fact that certain judgments *are* under-pinned by a plausible, albeit ideological super-structure which explains why they are accepted even when the evidence is lacking. Ideology itself (I am now referring to the overly-restrictive explanatory framework which shapes the specific judgment) is a structure which exists either to filter out or to transform what would otherwise be unacceptable experience. Quite simply, what *I mean* by 'ideological' is whatever habits of thought restrict our understanding rather than enhance it; whatever closes down areas, or aspects, of reality to the possibility of being experienced. One might say that the purpose of an ideology is to ensure that certain types of evidence drawn from experience *are* absent and therefore do not hinder us in reaching the conclusions we desire. Whereas the biased judgment must stand alone against a back-drop of beliefs with which it is ill-fitting, an erroneous judgment which is grounded in ideology is, in sharp contrast, surrounded by a tight defensive structure. It is protected against criticism unless that criticism is directed first towards the demolition of the protective structure itself. This is why I say that ideology is a greater threat to the expansion of understanding than is personal bias.

Nothing very much as been said up to this point about what I described at the beginning as the 'so far admittedly vague notion of *poor judgment*'. Now that we have examined the relationship between biased and (specific) ideological judgments, between ideologies (as whole structures) and the real world, we are, I hope, in a better position to elucidate the meaning of 'poor judgment'.

One point which has become clear is that, although ideology and bias can be sources of factual error, they need not be so. Whereas some forms of error (due to ignorance or misinformation) can be identified by showing that the subject's beliefs do not correspond to the *object* (the independent world), the sources of the specific errors we have been discussing in this chapter may be identified by looking at the *subject* who is making the knowledge claims. It is the *attitude* of the subject which distinguishes the merely ignorant from the biased or dogmatic individual. Specifically, it is the lack of integrity with which the subject goes about his or her enquiry into the nature of reality. This lack of integrity may be displayed either in the inconsistent application of a given interpretative framework (bias) or in the willing commitment to a restrictive framework *because it is perceived as such*, because the framework insulates the subject against the world which he or she has declined to know more fully.

The unwillingness to understand the world more fully has its origins in a stubborn commitment to certain values or aims whose maintenance requires that one's view of the world is not radically challenged. Thus there is a reversal of the proper order of priority (the order which is consistent with the attainment of objective knowledge) whereby one allows reality the upper hand in shaping one's values and goals. The latter must give way to the former where there is tension within an interpretative framework for the reason that the framework *exists* in order to expand our knowledge of reality. The framework must accommodate reality; reality should not be made to fit within an inflexible framework. Under the influence of ideology, it is one's values and goals which take the upper hand, determining the nature of the framework as if *they*, and not the world, are simply 'given' and must therefore be accommodated at all cost. When the order of priority is reversed, reality must be 'reconstructed' – and thereby distorted – to uphold the values which the individual holds most dear.

I have argued above that it is not the infiltration of subjective valuations *as such* into our explanatory structures which undermines their effectiveness. It is only the infiltration of certain kinds of valuation which is to be counted as illegitimate. I have called the ability to make the right valuations, and to bring them to bear upon our understanding of the world, 'sound judgment', and the converse, the tendency to give a significant role to intrusive or inappropriate valuations, 'poor judgment'. It is now possible to explain why and when a valuation plays an illegitimate part in theory-formation.

Those values are illegitimate that are not subject to revision in the light of experience, that are taken as 'given' and therefore place an absolute restriction on the ways in which reality can be interpreted. We demonstrate impoverished judgment wherever the values we embrace, and which form an integral part of our interpretative framework, hinder the expansion of the world that can fall within our experience. Valuations which have become entrenched to the extent that they 'take the upper hand' in the interpretative process are the basis for prejudice and faulty judgment. They intervene prematurely, or wholly inappropriately, in attempts to understand the world as it is so that the world-view is fashioned to support *them* rather than vice versa. When sound judgment is brought to bear, it is not that subjective valuations play no part, but rather that the part they play is conditional upon their enhancing the understanding, upon their extending the world which is known to us. And so there will be a constant inter-play between perceptual data, the theoretical structures which assimilate data and which determine our descriptions of what we perceive, and the value-systems which feed into these structures; each of the components puts a check, so to speak, on each of the others (although the components can never be decisively disentangled). Sound judgment belongs to those individuals who persist conscientiously with the cross-checking procedures between their experience and the elements of their belief-systems. No single component is regarded as absolutely immune to the possibility of further revision (or re-description in the case of sensory experience) but will remain as part of the belief structure only insofar as it as it does not hinder the expansion of the world which is known to us. Thus, sound judgment may be described as the ability to recognise which part of a network of beliefs, values and explanations needs re-examination if certain puzzles are to be solved and if certain facts are to be explained (or perhaps *discovered*) so that, ultimately, our understanding can be enhanced.

Poor judgment, on the other hand, results from the determination to regard some part of the belief structure as 'fixed' so that our ability to perceive, to explain and to understand is to that extent limited. We become *less* adaptable in our creative responses to problems because we have raised beyond the possibility of revision certain crucial elements of the belief structure. In particular, it is the individual's *value* judgments which can be raised up to this 'untouchable' status and, having reached so elevated a position, can begin to 'knock reality into shape'. It is ironic that what often protects value judgments from revision, and so bestows on them the power to re-shape and distort our vision of reality, is the belief that value judgments *as such* play no legitimate part in the accumulation of objective knowledge. So, in effect, the denial that values play *any* part in enhancing our understanding has led to their playing a most significant part in the formulation of scientific explanatory structures.

It is to scientific explanation that I would like to turn next, to consider the role of value judgments, the relation between ideology and science, and to pursue the argument that it is ideology which is the real threat to progress in science and not bias. Bias is a relatively easy target for criticism and it is one which has obscured the real enemy from view.

2. Value-neutrality as an Obstacle to Progress

In Chapter 4, I indicated that certain kinds of academic dispute, such as that between post-Keynesian and neoclassical economists, can never be resolved within the confines of an allegedly value-neutral enquiry because the rival positions, which have their origins in alternative interpretative frameworks (at least one of which may be 'restrictive' or 'ideological') may be both highly self-consistent and compatible with many of the widely-acknowledged facts. Thus I commented that it is very rare for anyone to change camps in the course of such a dispute; rather, the gains that are made on each side will take the form of achieving ever greater levels of internal consistency in response to the rival's attack, or in achieving a greater capacity for predicting or explaining phenomena (though of course the sides will disagree over the significance which is to be attached to the successful prediction). If anyone does shift camp it will not be because of some further empirical observation which s/he has previously ignored (like our referee above), but rather it will be because all of the *old* facts have been viewed from a new perspective; that is to say, a new framework for interpretation will have been brought to bear upon the already available data.

In Chapter 5 I suggested that no argument which purported to be value-neutral could persuade anyone to take up a radically different stance such that their old framework for interpreting experience is set aside and an alternative one, perhaps even that of their one-time rival, is adopted. Yet it is precisely this kind of switching and comparison of fundamental perspectives which is needed if, in the first instance, we are to understand one another and, secondly, if we are to have any hope of resolving such disputes. By insisting on the value-neutrality of their discussion, social scientists are simply ruling out of court the kinds of consideration which must be brought to bear if the discussion is to be fruitful. Indeed, they should be acknowledging just those kinds of value-based considerations which have already infiltrated their allegedly value-neutral theories but which have only muddied the waters of debate through their remaining unacknowledged. Unexamined and unquestioned they simply gain the upper

131

hand in the interpretative process. It is only by broadening the scope of the discussion in the ways I have suggested, and in recognising the role for certain types of imaginative or emotionally stimulating argument not usually admitted as 'rational' or appropriate to social science, that we can take on radically new perspectives, exchanging one interpretative framework for another so as to judge for ourselves which is the better (the more comprehensive) of the two.

Given an unpreparedness to make this shift, and the associated unwillingness to extend the scope and nature of one's arguments, then uncovering what one takes to be errors in another person's views becomes a most daunting and frustrating task. It may not be the case that a rival school of thought is simply *internally* inconsistent. It may simply be that the whole world-picture is what we wish to reject, even though the picture itself is reasonably coherent or could be made so with a more determined effort by its proponents. Our rejection of particular judgments or theories to which a rival approach (or world-view) gives rise then becomes difficult to explain to those who have internalised the world-view in question. Often arguments for one's own position involve the rejection of *specific* claims which appear to us as erroneous but which, as I have argued above, *can only appear erroneous* from the perspective of someone who already shares our world-view, or at least, to someone who has already rejected the framework for interpretation from which the specific (erroneous) judgment has arisen. Thus to address our arguments against specific claims in cases where the claim is actually consistent with the person's other beliefs may really be quite futile. The only way to break this stalemate, to demonstrate the error of any particular judgment, is first to demonstrate that the interpretative framework (which expresses the world-view) is itself erroneous. If the framework is not challenged then the supposed 'facts' will remain firmly encased within the protective structure of ideology.

An interpretative framework as a whole may be said to be erroneous in that it results from a partial or restricted involvement with reality; a withdrawal of the knowing subject from some aspect, or aspects, of the world. The error, in such cases, consists *not* in our being ignorant of certain facts, nor in our acknowledging facts sporadically when it suits us, but in our being unable to recognise them at all, in our being closed to the possibility of having certain experiences, of experiencing the world in all the ways in which it can be experienced. This is the world-diminishing effect of labouring under an ideology. Properly speaking, it is the interpretative framework only which can be said to be 'ideological', although it is particular judgments which provide a first clue to the nature and very existence of a restrictive, or ideologically distorting, framework.

I have already remarked in the previous section that one particular kind of error, bias, is in principle at least, relatively easy to eliminate. I say *in principle* here only because I recognise that people's obstinacy can in practice make the elimination of any error impossible, however conspicuous that error may be. Such extremes of obstinacy may, in fact, be explicable only in terms of the wish to defend, against actual or merely perceived threat, an essentially ideological world-view. In claiming that bias is in principle more readily eliminated than those errors which are due to ideology, I simply mean to draw to attention the inherently insecure position of the biased judgment within the network of beliefs currently held by an individual. Given the existent framework for interpretation, recognition of an alternative non-biased judgment ought to be possible. The relevant kinds of experience which will overturn the biased judgment are perfectly accessible, indeed, by definition they are accessible, for if the initial judgment is truly biased – and not merely an expression of ignorance for example – then what this *means* is that the facts are already known, or straightforwardly capable of being known, but are being either over-ruled or simply ignored under the influence of some prematurely intervening valuation. The biased individual 'looks away' from what s/he would otherwise see as facts.

In contrast, when labouring under some form of ideology, there is often no need to 'look away' from the facts which others would have us acknowledge: given that our existing framework for interpretation is left unchallenged, the facts which others may be urging us to see simply do not confront us, or they do not strike us as relevant in the way that others perceive them to be – it is not that the facts are perceived *differently* but that *different* 'facts' are being perceived. It is for this reason that I say errors resulting from ideologically distorting frameworks are so difficult to unmask. They are revealed in most cases not by looking at new evidence, but by looking in a new way, perhaps more profoundly, more sympathetically, at the most familiar areas of our experience. Arguments which are supposed to be value-neutral cannot produce the sort of fundamental re-appraisal of a person's beliefs that is required to unmask an ideology.

There are, of course, ways other than argument of bringing about dramatic realisations of past error. Suddenly being brought face to face with shocking, yet unfamiliar, circumstances may cause an individual to alter many of his/her old beliefs and values. For example, the sheltered middle classes of the 1930s, confronted with Orwell's harrowing account of working class life in *The Road to Wigan Pier*, were forced to look again at many of their most cherished beliefs about British society. Catholic missionaries too found that their perceptions of the USA were much altered once they had lived and worked amongst the people of Cuba.

These particular examples bring out a further point: when we expand our circle of those for whom we feel sympathy, we are able to expand our theoretical understanding of how their situation relates to our own, of how one society, or group within society, affects and in turn is affected by others. Our understanding of life in Britain in the nineteenth century is limited if we do not see it as at least in part sustained by a much lower standard of living in the British colonies. Similarly, we cannot explain or properly comprehend all of the features of white South African society without looking beyond that society to the black South African shanty towns. Any attempt to furnish an explanation without seeing or wishing to see the relevance of looking any further could only produce a profoundly distorted picture. I would suggest that the more all-embracing one's sympathies become, the more one is able to provide a full and satisfying account of social and economic phenomena.

We can see from these examples that our interpretative framework may be amended not only by a deepened understanding of our everyday experiences, but also by a completely new, previously inaccessible experience. The point I wish to stress here is that there are certain kinds of error which can only be cleared up by achieving an expansion of one's world, either by penetrating more deeply beneath the surface of everyday life, or by extending the boundaries of our world to embrace what lies beyond our current experience.[2] If we fail, or refuse, to make such an expansion then, with regards to the limited world we have chosen to inhabit, we may yet subscribe to a set of reasonably coherent, largely consistent and even truthful beliefs. So long as no-one else's world (or *the* world) impinges upon our own we have no reason to think that we may be in error. Unless we are shown that there *is* a world beyond are own, unless we are compelled to confront it, *to get personally involved with it,* by means of a more graphic or imaginatively stimulating argument, then no-one will convince us of our error and their arguments will remain futile.

It should be clearer by now why it is that I have spoken of ideological frameworks as 'restrictive' and 'distorting'. An ideological world-view may well be – indeed, is very likely to be – quite coherent and, to a very great extent, successful in guiding individuals' activities towards their chosen goals. Obviously, if it were not so, it would be hard to explain how anyone could come to be in the grip of an ideology. It would be an embarrassment for those of us wishing to take the 'false consciousness' interpretation of ideology to be apparently committed to such a dim view of human rationality, one which seemingly must depict so much of humanity as steeped in irrational prejudice and therefore incapable of reasoning in a coherent fashion to dispel their false beliefs. Though claiming that the ideological world-view is in a certain sense false, I have not been claiming that every single proposition or valuation which expresses an ideology is either false or misguided. On the contrary, for an ideology to have any persuasive strength (and undoubtedly they do have such strength) it must provide an apparently adequate interpretation of much that lies within an individual's personal experience. The sense in which the choice of interpretative framework may be erroneous is in so far as it impedes a fuller involvement with, or understanding of, the world as it is. The ideological view mistakes a merely partial, and to some extent therefore, distorted view of the world, for reality *as a whole*. Under the influence of ideology we choose (perhaps with some encouragement from society or our peers) to draw in the boundaries of our world so that *the* world is not identical with the world of our experience. Our vision or perceptive power becomes restricted; we seek not to *encompass* reality as a whole, to affirm as much as we may, but rather we tend towards denial, or, where denial is not possible, towards the systematic distortion of those features of society (or of the natural world) which do not conform to the idiosyncratic world-view. Thus, the effect of ideology upon knowledge is not to render it all false: much of the world is readily and accurately comprehended. What occurs is an impoverishment rather than an invalidation of one's personal experience, an alienation of the individual from that which might be known but which is left unknown, or misconstrued, in an effort to gain that sense of security which accompanies the retreat into even the most partial isolation from the world around us.

In effect then, what the ideologically distorting framework succeeds in doing (indeed what it *exists* to do) is to filter out certain subjectively unpalatable aspects of the real world. These may be few or many, but even for the most insane it is unlikely to be all and every aspect of the world which is rejected. A filter so complete would no longer be a filter but an opaque screen, and in any case, the result of such persistent and thorough-going error regarding the nature of the world would be, without doubt, the most unpalatable existence of all.

In defining ideology as that which restricts the contents, and therefore the understanding of one's world, I might seem to be placing myself in the awkward position of claiming that *everyone*, without exception, is under the influence of an ideology. Granted that none of us is omniscient, it follows that we must all have a restricted or partial view of the world and hence, apparently, an ideologically distorted view. Fortunately, this is not an implication to which I am committed. The partial or incomplete knowledge of the world which, as finite beings, we cannot but help possess, need not be 'ideological' for the reason that I have not been using the term 'ideology' as if it were synonymous with unavoidable ignorance. I restrict its usage to a species of ignorance only, *to those types of error which could be avoided if certain choices were made.* In defining ideology, and in discussing its relation to the concept of bias, I have been concerned to stress that both ideology and bias result from a lack of integrity on the part of the enquiring subject. Only those choices which are made *without* integrity lead to the types of error of which I have been speaking. Ideology, specifically, is the product of an *alienated* choice. Of course, no choice of ours could overcome the limitations of a finite intelligence but there are choices we could make which would expand our understanding beyond what it is at present. What *does* distinguish ideology from straight-forward ignorance – due to mental limitation or the unattainability of information – is the unwillingness of those under its influence to incorporate all that they might (given their abilities, opportunities, and available data) into their subjectively apprehended worlds. Such unwillingness is itself demonstrated in the very reluctance of which I have spoken to take up another perspective and to discover for oneself what may be seen from another's vantage point.

I have argued in Chapter 6 (section 2) that we do have grounds for comparing and making a rational choice between competing theories and it may well be that having performed such an exercise we find ourselves back where we started, with the world-view we were inclined towards in the first place. Even such an exercise is not without value however, for it results from a healthy realisation of the possibility of our having been in error. It is *because* we are finite beings that we can never afford to behave as if we have got our enquiry over and done with, to behave as if we have found the answers, *know* them to be the answers, and need therefore search no further.

A closed and complacent mind – one that thinks it has got all its learning over with – turns all that it knows into mere dogma. For such individuals, beliefs are not maintained because their truth is being continually re-confirmed but rather because their truth is no longer put to the test. The world-picture which is perceived as 'finished' by its creator is one which is likely to be self-contained (coherent without the need for further information). However, the creator is also likely to be most unwilling to make any further addition or alteration to the picture which is taken *as it is* to be accurate and life-like. Most importantly, the debate about further improvements to the picture is closed and any suggestion that it is not provokes the kind of animosity which usually betrays the most deep-felt insecurity.

On the other hand, the mind which is open, restless and inquisitive is characteristic of the individual who fully recognises his or her own limitations and so remains alert to the possibility of error. It is this attitude which transforms a set of beliefs which may, in part, be *factually* mistaken (given that we are, after all, finite) into a world-view which is essentially non-dogmatic and non-ideological in nature. Such a mind cannot be accused of arrogantly asserting that adherents to defunct theories of the physical universe, were necessarily labouring under some form of ideology simply because *from our perspective* we can see that they had a limited world-view. They might have laboured under an ideology from the point where they ceased to be rational adherents to one theory and became opponents to a newly emerging theory *without even grasping its merits*, but so long as the older theory was actually the best theory on offer – and had not, for its adherents, acquired the status of unquestionable truth – then they could not be said to have been the victims of ideology but only of the technical limitations and general ignorance of their time. Doubtless from the point of view of humanity a hundred years from now we too will appear comparatively ignorant but, it is hoped, not guilty without exception of being dogmatic, un-self-critical and unreceptive to new ideas, for if this were the case then we as a society will have added to the already sufficient handicap of finite and fallible intelligence, the extra burden of ideology.

3. Science and Ideology

In his paper 'How to Defend Society against Science'[3] Feyerabend presents
the view that science is not exclusively or necessarily the champion of
society as regards the revelation of truths and the banishment of ideology-
based errors and human ignorance. Whilst it is undeniably true that
scientists were at the forefront of the attack upon superstition and upon the
stultifying authority of the Church, as science itself has progressed, so its
role in society has altered. Feyerabend remarks that where science had once
served to liberate thoughts, to expand the boundaries of the known universe,
now it has begun to take over the very role for which scientists of old had
criticised the Church: that of sole authority on what is to be taken as 'truth',
tolerating no contenders for this position, and final authority on the proper
methods for attaining such truths. Thus, in modern society, science has
become the tyrant. The fact that it was once a liberator from the ideologies
of a bygone age does not in any sense demonstrate that it cannot have
become an oppressive ideology now, for there is nothing inherent in science
which makes it *essentially* liberating.

Feyerabend's arguments are not of course directed against science as such
but rather, against scientific claims – and claims made *for* science – which
go unchallenged, and against those scientists who consider that they have no
serious rivals outside of the scientific tradition. The lack of critical
opposition both from within the scientific community and society at large
transforms the body of scientific knowledge into mere dogma. The debate
between science and non-science has long since been closed for many
individuals; it is felt that any further *valuable* contribution to our knowledge
will most certainly come from those who have painstakingly followed
scientific procedures.

The firmly established truth which has, for some, been elevated beyond the
possibility of further doubt, is that scientific method is the *only* appropriate
method for extending our knowledge of the physical world. Yet in their
unshakeable commitment to this belief, scientists have unwittingly laid the
foundation for the impoverishment of their knowledge. This absolute
commitment renders all the individual's other beliefs, whether true or false in
their own right, as elements of an ideology, because what the belief in
question signifies is the 'closing off' of possible, future channels of enquiry.
The only world which can then be known is the world accessible to science,
and this may not be *all* there is to know about the world we actually inhabit.

Feyerabend, I think rightly, suggests that science itself may become an oppressive ideology.[4]   However, unlike the religious, metaphysical ideologies which it has displaced, science has at its disposal more effective weapons to combat criticisms of this kind. I have already argued that progress in science, as in any area of human enquiry, requires that we lift the restrictions on what is up for debate. So, when the faculties for judgment are sound, there will be a constant cross-checking and re-examination of *all* the different components of one's belief system, including reports of one's perceptual experience, the theoretical explanations given for these experiences, their relative significance and implications for action, as well as the values, goals and preferences which help shape the explanatory structure itself. When some part of the belief system is raised above criticism then our ability to comprehend and explain the world we live in is diminished; our belief system is rendered less adaptable and we are less able to cope with surprising or anomalous occurrences. Science in particular has been able to defend itself (with some degree of success) against the accusation of being 'just another ideology' because, unlike previous metaphysical or theological ideologies, the dominant image of science is that of a *value-free* systematic enquiry into the nature of the universe. So the scientist is *not* vouching for the objectivity of any particular set of values and is *not* commonly perceived as either relying upon, or strictly implying, any specific valuation. In other words, the value component is not rigid and beyond criticism: it simply has no place in scientific investigation. It is left to float free but *outside* the confines of the scientist's laboratory. Of course, science may be used to serve various ends, some of which may be reprehensible, others admirable, but this does not affect the claimed objectivity of the scientist's research findings. The idea is that once we have obtained an objective account of the natural world, we must then decide what to do with it. The scientist *qua scientist* cannot help us here because such decisions are ultimately value-determined and therefore subjective – and the scientist is concerned only with the *objective*.

In stark contrast, the theological ideologies which, in the Western world, science has now largely superseded, explicitly included sets of values for which the ideologies' proponents sought universal acceptance. The inflexibility of the value-component in their belief systems was thus painfully apparent and provided a fixed target for their critics, particularly those, like Galileo, who wished to expand their factual or theoretical knowledge of the universe but found the value system too restrictive.

Science was able to supersede such ideologies, and to perform the liberating function which Feyerabend attributes to the early days of scientific advance, precisely because the view of the world which science presents – objective, independent and value-*neutral* – is one which undermines all explicit value systems equally. Preference for one value system over another could not be justified scientifically, given what science purported (and was widely perceived) to be, so some other non-scientific justification had to be found to support the values which religious authorities had traditionally proclaimed to be universally valid. Then, as science began to claim for itself the monopoly over objective knowledge, everything that was left outside the boundaries of legitimate scientific concern was, by implication, found to be lacking the type of justification required to secure *universal* acceptance.

Thus theological ideologies were gradually undermined by the increasing respect for the achievements of rigorous scientific method. The values which had been imposed on the ordinary individual 'from above' were, in time, recognised *as* impositions because their acceptance could not now be argued for on the grounds of their objectivity: they merely reflected the attitudes and interests of the dominant social group. The ordinary individual, meanwhile, had other interests and so, it seemed, might well choose to embrace another set of values which better expressed those interests. In any case, neither the scientist nor any other 'expert' could lay down a moral code and still claim its validity for all. In this way, science was indeed liberating. It unmasked moral codes which had purported to be valid for all time and for all people and revealed them to be the instruments of oppression. At least, it was *science* which awakened the critical faculties to this alarming possibility and, once this was achieved, the foundation was in place for the linking of moral and political relativism with such positive notions as toleration and freedom of expression. Science thus gave the impetus to social reform and feudal society was edged a step closer to the political modern age.

By comparing science with the theological ideologies which it has displaced we can come to see how science may defend itself against the accusation of being simply another ideology and one which, with the passage of time, has become just as repressive as those it has replaced. It may be argued that, since true science is value-neutral, it cannot be the case that the scientific community, or a society founded on genuine scientific principles, is guilty of imposing any particular value-system on the populace. After all, the scientist explicitly denies having anything to say on questions of value preference, on questions of what are the most desirable social goals, on how we should arrange our lives, or on questions concerning behavioural norms. These are all questions which concerned the Church in one way or another but *not* the scientist. Scientists apparently confine themselves to advising us how best we may achieve whatever goals we have freely fixed upon, but no more than this. How then could they be accused of being party to an on-going, systematic repression? It would appear on the contrary that the more reliable and extensive knowledge of the world which has been provided for us by the scientific community has actually enhanced our freedom. Science, it is argued, has delivered us for all time from the oppressive, ideological belief systems of the past insofar as it gives no grounds for a preference (or an imposed preference) for one set of values or social norms over any other. Science, therefore, may present itself to the public as that which dispels ideology and allows us to see the world as it really is, without any subject-imposed, value-based distortions.

The conception of science being opposed to ideology, as truth is opposed to error, is one which has proved hugely influential in Western thought. It has helped shape both the public's perceptions of science and many scientists' perceptions of their own activities. This conception of science as being the only means for actually overcoming ideology, as involving a distinct form of cognition from that which produces merely *ideological* belief, is a conception which finds expression in the works of Auguste Comte. In his 'Cours de philosophie positive',[5] Comte describes what he sees as the three stages through which the history of human thought has passed: the theological or fictitious, the metaphysical or abstract, and finally, the scientific or positive phase. The first two are distinct from the last of these phases in that both the theological and metaphysical outlooks share a 'predominance of imagination over observation'. One might equally say: 'a predominance of the subjective over the objective' but, in any case, the intended contrast between the scientific approach and the others is clear. The subject's contribution to understanding – the imagination – has, according to Comte, left society in disarray. Again in 'Les Cours', he expresses the belief 'that ideas govern the world, *or throw it into chaos* – in other words, that all social mechanism rests upon *opinions*.' The goal of positive philosophy is thus to weaken the influence of harmful opinion (ideas or 'ideology', though this is not a term Comte uses) and to bring about a rational social order founded on the objective observations of social scientists. Comte sees it as his task to bring social science into the 'positive' phase of its history, alongside the more developed natural sciences.

Comte himself has been criticised for not being true to positivist principles – many of his own assumptions owe more to metaphysics, to the imagination, than to observation – but this inconsistency has not diminished the appeal of his central ideas. What has endured, and re-emerged in the *logical* positivism of the Vienna Circle, is the image of science as essentially in opposition to metaphysics, to the products of the human imaginative and speculative faculties which merely serve to distort our vision of reality. The concept of ideology which has evolved from this particular view is one which equates ideology with arbitrary (empirically unfounded) speculation, speculation which cannot be elucidated in terms of anything observable and must therefore be judged as meaningless. In short, 'ideology' may be understood to refer to the meaningless products of human imagination and since science is the product of human *observation* (coming from the object and not the subject) it *cannot* be ideological.

In this century, the most effective response to the accusation of being ideological has been the counter-attack. This begins with the appropriation of the term 'ideology', followed by its re-definition so that it is no longer applicable to what one has defined as 'science'. The self-styled image of the scientist as value-neutral observer has made this mode of counter-attack possible. It is because science has claimed this special status, that of being cut off from subjective valuations which infiltrate other types of knowledge, that it can present itself to the individual as the only real alternative to – or liberator from – ideological belief systems (where the latter are understood to rely upon, or incorporate, the subjective valuations which the scientist shuns). But I have argued already that no-one can be a scientist in this sense for the reason that theoretical, explanatory structures do not come together of their own accord, that is to say, they are not compelled to come together in a certain way by the 'facts' alone, but are *put* together by an enquiring subject who thereby plays an active part in gaining a better understanding of the world. This contrasts sharply with the value-neutral view of science which assigns to the scientist a more passive role: if the individual puts in an appearance at all (makes a subjective contribution) then the result is taken to be a less than *objective*, a less than *scientific* and truthful account of reality.

I have suggested that, in defence of science, the most effective response has been to appropriate and re-define 'ideology' in such a way as to exclude the possibility of its applying to a thing like science, given the way 'science' is defined. I am disputing (a) the use of 'ideology' to refer unfavourably to any belief systems that are developed with the aid of an active, creative contribution from the knowing subject, and (b) the view of science (or, at least, *good* science) as an essentially value-neutral enquiry. Here, I wish to concentrate on (a) the use, or meaning, of 'ideology'.

In this section we have considered the role of science as liberator, releasing ordinary individuals from their perceived obligation to assent to the values of, for example, the Church. But we have not yet considered *why* the intellectual and moral liberation of the individual might be considered important. It cannot be, from the point of view of value-neutral science, that the individual's freedom of thought and decision is valuable *as such*; rather, it must be that the possibility of science is seen to require the belief that values are fundamentally different to facts, and different in such a way as to make it imperative to scientific progress that they are not allowed to interfere with the investigation of facts. So fact and value must be kept apart, and one way to do this is to promote the perception of values as arbitrary, non-rational, as coming *from the subject*, whilst facts come to us *from the object* and therefore constitute knowledge of something which is both external and independent of the will. With the basis for objective value-choice undermined, it really does not matter (to science) what ultimate choices individuals or societies make so long as these choices do not influence and impede scientific progress. Thus it seems that all that freedom can mean from the perspective of a supposedly value-neutral science is the freedom to make arbitrary choices and the freedom to reject as valid for ourselves the arbitrary choices of others which might be imposed upon us. Meanwhile the scientist *qua scientist* is apparently *not* exercising any freedom of choice, except insofar as s/he chooses whether to engage in science, but instead leaves all power of determination to the neutral, observable data.

What the liberation achieved by science amounts to is this: the power to choose and inculcate values is wrested from the hands of authorities who had used the power to sustain their position, and placed in the hands of the individual – but only after the power to choose is rendered practically worthless! Obviously, we are not talking about material power here, of either the political or economic varieties, but about the *feeling* of empowerment which accompanies any form of enlightenment. The recognition that one need not accept the values of those in authority, as well as the realisation that the *raison d'être* of the value-system is to serve the interests of those in authority, initially produces a feeling of *personal* liberation. If the scientific presupposition concerning the relative status of facts and values is internalised then liberation is achieved in the relevant sense. If it is universally internalised then the old-style authority is undermined and the political and economic power structures undergo a transformation which reflects the change in public consciousness. (Historically, this change corresponds to the gradual evolution towards modern liberal democracy.)

If we understand by 'ideology' those habits of thought which place limits on the world which is accessible to the subject, then it is clear that the value-neutral conception of science, if it is internalised, does limit what the subject feels able to achieve. Ordinary laymen cease to develop as enquiring subjects because they cease to expand the world that is personally known to them. The acquisition of objective knowledge is, by and large, left to the scientist and little remains for the non-scientifically trained to learn about reality. Much of what does remain suffers an 'epistemological demotion' and, by implication, those who can only concern themselves with these remnants have their status undermined also. The result is that the ordinary individual has no more elevated a status now than under the non-scientific, metaphysical or theological ideologies of the past. The main difference is that the ways in which the individual is undermined are far more subtle.

The enquiring subject whose critical faculties were once primarily under attack from without (from the Church or State seeking to impose an explicit and rigid value-system) is now under attack from within. Having been convinced – partially by the superior explanatory power of science, partially by its associations with a more prosperous, more tolerant society – that the active participation of the subject is detrimental to the acquisition of objective knowledge, the subject is all the more easily pacified. Science is left untouched by the value-inspired criticisms of the more rebellious individuals because it apparently leaves *their* values untouched. Science as an ideology is infinitely more accommodating than its ideological predecessors and this is the reason for its great appeal. Whatever one's opinions, one is inclined to turn to science for support, to find the materials for attack upon those with conflicting opinions, whilst science itself, as the value-neutral arbiter in such cases, usually escapes attack. It rises above disputes in a way the Church, for instance, could never achieve.

My main point, however, is that in the relevant sense science is akin to other belief systems which we may call 'ideological' in that it too has contributed to the inactivity of the subject in what must be a *personal* search for knowledge. (I argued this in Chapter 2, section 2, 'The Individual and Claims to Knowledge'.) In one way or another, the full use of the individual's critical and creative faculties are discouraged, either by means of force, by ex-communication or public derision, or else by more effective means – by persuading individuals that it is in their interest, in upholding a set of practices for obtaining a truly objective body of knowledge, to keep a tight rein on their subjective impulses. For those of us who are not sufficiently dispassionate to make good scientists, a more extensive knowledge of the world must always, apparently, belong to somebody else.

Notes

1. To clarify, if we wish to claim that a pro-monarchist stance is ideological then what this claim *means* is that a person can only acknowledge the authority of a monarch on the proviso that aspects of reality are either systematically ignored or misrepresented. Thus, for the anti-monarchist, to grasp the fuller picture of reality is to see that, on balance, the presence of the monarchy does more harm than good.
2. A restricted world-view (ie. limited by an acquaintance with only a sub-section of society or of the international community) can only produce theories of limited explanatory power because to understand society properly we have always to look beyond its defined limits.
3. Feyerabend, P., *Science in a Free Society*, New Left Books, London, 1978.
4. He does not use 'ideology' as I do. For Feyerbend *all belief* systems are ideological: they are 'fairy tales' that contain much of interest and value but which also contain lies. I maintain that it is at least possible to have a non-ideological belief system.
5. Comte, A., 'Plan of the scientific operations necessary for reorganising society', (Third Essay 1822), in G. Lenzer (ed.) *Auguste Comte and Positivism, the Essential Writings*, Harper and Row, New York, 1975.

8 Today's Ideology

1. Science, Modern Liberalism and the Market Mechanism

In Chapter 7, I argued that a so-called 'value-neutral' science could not *as such* raise itself above the accusation of being ideological and, furthermore, that there are important ways in which this very popular conception of science forms an integral part of what is arguably the most insidious and criticism-resistant of world-limiting ideologies. The ideological belief system to which I am referring here is that which characterises the modern liberal,[1] market economies of the technologically advanced Western world. I suggest that it is no coincidence that value-neutral science is held in high regard in precisely those cultures (though not only those cultures) which either are, or are evolving towards, a condition which is often with unquestioning approval described as 'liberal democracy'. Of course there are many benefits to living in *this* sort of society: I do not for a moment think that life in Britain would come off unfavourably from a general comparison with, for instance, life in Cambodia or the old Soviet Union. It is just that I feel an alarm should be raised when it is taken for granted that *just one particular form* of social organisation (which liberal democracy is, one amongst a number of possibilities) is constantly assumed to be the final and most nearly perfect stage of our long and turbulent history of socio-political evolution. When another country expresses the desire to join us in this state of organisational near-perfection we feel encouraged by the way the news is presented to let out a cheer: yet another one of our evolutionary running mates has crossed the finish line. There may of course be problems with the new system – we all have those, even the most well-established of democracies – but the basic, most common-place conviction seems to be that a liberal democracy will provide the best framework within which to deal with any problems. It is this conviction, held on to by many in the face of very serious, evidently widespread social and psychological problems,[2] which gives us our first clue to the existence of a world-limiting, alienating

146

ideology, and in what follows I would like, amongst other things, to give further clarification of the place which a *value-neutral* conception of science holds within this ideology.

Before continuing, I should say that the ideology to which I am referring, like all ideologies which we attribute to entities like 'states', 'societies', 'classes' or 'sects', is not an easily definable thing. That is to say we cannot pick out all the specific beliefs that one must hold in order to be said to subscribe to an ideology-*type*. Aside from using the term to refer to the belief system of a single individual (which is how I have mostly been using the term 'ideology' up until now), the term may also be used to group together large numbers of people and their beliefs, brushing over individual variations. What I have been alluding to so far as the ideology of the modern liberal, market economies of the technologically advanced Western world is a wordy, though necessarily loose description of a belief system which I hope the reader has a sense of in much the same way as one can grasp what is involved in belonging to, for instance, the Baroque tradition in music.

For convenience, I had better find a less wordy, though perhaps overly simplistic, label for the ideology I wish to discuss. Let us call it the 'Capitalist Ideology' and make our clarifications here concerning what sort of beliefs it might embrace. Firstly, it is, as I have said, an ideology-*type*, which is to say that many specific ideologies (or *individuals'* alienating belief systems) may fall within this category, though the individual belief systems themselves may differ in their detail in a wide variety of ways. Secondly, though 'capitalist' refers strictly to a mode of economic organisation there is a precedence established by Marx for using the term to refer broadly to the current period in European and American history, to the many inter-related aspects of life within this period. Marx speaks of a 'capitalist ideology', of the 'ideological superstructure' of capitalist society and its relation to the economic base, and with Marx I would agree that there *are* intricate links between the mode of economic organisation, the mode of political organisation, and many of the features of the dominant belief system, so that one may usefully characterise them as essentially 'capitalist'. Some of the beliefs which would comfortably hold a place within the capitalist ideology include the following: the efficiency of the market to regulate the distribution of goods; the desirability of perpetual economic growth; liberal democracy providing the best means, indeed, the *only* means, for representing interests and resolving disputes; the benefits of technology and the importance of science; toleration; freedom of speech. Hopefully the 'flavour' of this ideology has been conveyed by what is, of necessity, an incomplete list of the beliefs held most dear within the culture of twentieth century capitalism.

I have pointed out that what makes an ideology so secure once it has got a grip, is that characteristically it possesses a very high degree of internal coherence and at least some aspects which are almost immediately appealing. Not *every* judgment or belief within an ideology is mistaken, only some are, and those that are true combine with those that are false in such a way that the latter are secure from an easy refutation. Thus, in saying that, for example, a belief in freedom of speech may 'comfortably hold a place within the capitalist ideology' I would not want to be construed as saying that freedom of speech is therefore an undesirable thing. What I am trying to do is to draw to attention a complex belief system, to understand how a particular set of beliefs knit together to form what is undoubtedly a very coherent and appealing belief system, but one which is nevertheless ideological because *some* of those beliefs close off the possibility of a fuller involvement with reality. The beliefs which are alienating are, in the most successful ideologies, anchored in to the belief system through combining with some of our most deeply felt convictions – like the belief in freedom of speech – so that if one attacks the more controversial assumptions one is seen to be undermining what is held most dear. The problem for critics then is to disentangle and remove what is wrong whilst retaining what is essentially correct.

Capitalist ideology appears today to be peculiarly criticism-resistant – whilst other ideologies have crumbled or fight hard to maintain their grip, it keeps a tight hold and indeed is spreading to fill the ideological spaces left by the less compelling rivals.[3] The leading representatives of the capitalist ideology have not needed, at least in the twentieth century, to take very seriously the threat of revolution on their home ground as have their counterparts in other, ideologically distinct, régimes. The main reason for this, I suggest, is that they have far more subtle means at their disposal for maintaining the economic and political status quo, even in periods of economic hardship.

It was the capacity to maintain stability and to harness, even to re-direct, the rebellious elements which capitalism provokes, which Marx could not have anticipated. He could not have foreseen the durability of the capitalist mode of organisation because (a) at the time he was writing, the capitalist ideology had not evolved into the highly criticism-resistant form it has today (where it is more consistently linked to modern rather than classical liberalism) and (b) Marx himself had not identified all of the elements of the ideology and remained in the grip of an alienating view of science which was to ultimately undermine his critique.

An important part of the ideology is its built-in safe-guards: a conception of rationality and of what is required for objectivity (which is both reflected in, and perpetuated by, the value-neutral conception of science) which in effect cripples the critical faculties of those who might otherwise throw light upon the causes of society's malaise. Much more will be said in what follows of the links between a certain conception of rationality and of objectivity, alienation, and life in the capitalist world as I attempt to draw together the main elements of this work. However, I will begin with an examination of what is arguably the most seductive aspect of the capitalist ideology: its emphasis on freedom.

'Freedom', and the related concepts of 'democracy' and 'choice', are the key words that must be employed in any successful piece of political persuasion in both Europe and the United States. It is imperative that these words are associated in the mind of the public with the government's every action because they effectively summarise the form that any acceptable justification for policy must take. Even the critics within such societies invoke these key words in appraising government policy, saying that it is, in fact, not consistent with individual freedom or with democracy, but such criticisms cannot be – and indeed are not intended to be – radical criticisms of the political system as such. The ideals of capitalism are not generally challenged: a question has simply been raised as to whether the government has actually lived up to them, and it remains the case that the framework within which most political debate takes place is the liberal democratic one. The participants share, or must claim to share, common ideals and have a broadly similar interpretation of what is consistent with the fulfilment of them. This latter point is important because of the possibility of sharing the explicit ideals of liberal democracy whilst denying other people's interpretation of what is required to satisfy these ideals. This, for instance, would be the anarchist's distinctive position in the debate.[4]

Within the liberal democratic/capitalist framework however, freedom is interpreted in such a way that it *is* seen to be consistent, or at least in principle reconcilable, with many of the features of society as it now stands. It is this perception which removes the main incentive to engage in radical criticism of the status quo. The means of control in other types of society, such as Iraq or Iran for instance, seem so very transparent, oppressive and threatening, that our relief at being spared the most crude forms of repression has perhaps blunted our awareness of the restraints that act upon us here. It is a unique feature of Western governments that the means of control which they (either consciously or unconsciously) rely upon are not *experienced* as a means of control by the vast bulk of the populace. Much of our language suggests that the mechanisms which regulate our lives, limiting the horizons of the individual, are perceived as being the very mechanisms *by virtue of which* we proudly see ourselves as members of 'The Free World'. Thus, being a 'free country' has come to be viewed as synonymous with being a Western-style democracy, and, in turn, being a democracy has come to mean having, to a greater or lesser extent, an unregulated market economy. How often have we heard recently that the Russians are now 'free', and even 'democratic', apparently because they are governed by market forces, because the wealthier minority among them are now driven by the profit-seeking motive? The terms 'market economy' and 'liberal democracy' seem to blend together at some point in the minds of journalists reporting on foreign affairs, and indeed they are not entirely mistaken in apparently using the terms inter-changeably. Journalists have a sense of the idea that 'market economy' and 'liberal democracy' refer to the two sides of the *same* coin, one referring to the economic, and the other to the political aspects of the capitalist ideology. The ease with which these conceptually distinct items are run together – human freedom, the market economy and liberal democracy – is certainly alarming but they seem so naturally bound together because of the way the belief system under capitalism is structured. Once the ideology has been internalised, it is anything but easy to disentangle these items and to raise serious questions such as: To what extent *are* we free in the so-called democratic, market economies of the West? And, in any case, what satisfaction does the freedom we are allowed actually give us? Does the political system really empower us, or is it possible that the Western governments have the advantage of an ideology which merely convinces us we are 'having our say', or let us have our say to a small and insignificant extent, in order to more easily pacify and control us? These sorts of questions are, of course, intelligible but generally not taken very seriously by the public at large.

Why should it be the case that the general public is so easily convinced it is really free, or as free as it is possible to be in a large, organised community?

The answer lies partially in the success of Western governments in the exercise of persuasive definition.[5] The term 'democratic' has been adopted as a self-description by governments – with some justification undoubtedly, for the modern form of political organisation is indeed *more* democratic than many historically previous forms – and the word has come finally to be identified with just the kind of political set-up we see today. But 'democracy' has an evaluative as well as a descriptive component so that once we agree that 'democracy' refers to something in the world and that, if it is actualised anywhere (however imperfectly) it is in countries like Britain, America or France, then it follows that the political systems which now go by this name are basically deserving of our approval: that is to say, they acquire all the positive connotations that go with a term like 'democracy'. Seeing as the term 'democracy' actually means 'government by the people' – in other words, the *empowerment* of the people – it begins to look necessarily absurd, once we have agreed to call something a democracy, even to ask the question: 'But does this *particular* democracy really empower us?' What better way could there be of off-setting the criticism that one is disempowering the populace in some fundamental way than to first of all appropriate the term 'democracy' for oneself, achieving universal acceptance of this self-description, and thus to secure the analytical truth of the claim that one's government is both the true expression and protector of the people's freedom of choice? To attack capitalist governments *without* denying that they have the right to call themselves democratic – except in a comparison with absolute tyrannies – can then be construed as tantamount to attacking the ideal of freedom itself. In short, to attack the government is to attack the people whom it claims to represent and so its radical critics (those who question the framework of political democracy itself) find themselves severely wrong-footed in the political debate. Acceptance of the current political system is bound tightly into the capitalist network of beliefs by virtue of the links that have been forged with the appealing ideals of individual freedom of choice and self-determination.

No successful ideology, as already pointed out, could be constituted entirely of beliefs that are mistaken. An ideology must explain, and be consistent with, a very large part of our experience in order for it to have persuasive power. If people were really living under a wholly oppressive régime they could not under any circumstances be convinced that they were free – but people *do* believe that this is a free country, and often say so, therefore it seems to be true that they are *not* living under a wholly oppressive régime, that they are, to some extent at least, free and self-determining. If this is correct then we have found the basis in reality upon which the capitalist ideology has staked its claim to the ideals of freedom and self-determination. In reality, the capitalist system actually does allow *more* freedom of choice; public opinion actually does affect some government decisions as well as (in very general terms) the political make-up of the government itself. Thus, we can very easily see where the grain of truth is within this compelling system of beliefs. The claim, for instance, that the market promotes diversity and thus greater freedom of choice *is* borne out by much of our day to day experience and is illustrated particularly well by any comparison with countries that have restrained their market forces and have also experienced great economic hardship. Our belief system thus receives regular confirmation from sources both at home and abroad.

We can, it seems, easily find some merit in the capitalist ideology. We can see the ways in which it *works* for its proponents. So where does it start to go wrong; in what ways might we be carried along by it further than the truth will allow? What seems to me to be revealing is that when describing the benefits of life in the capitalist West, we almost invariably use the language of comparison and qualification. We commonly make our point by contrasting life here with life elsewhere. Thus, we have 'more' freedom of choice and can affect 'some' government decisions; strictly then we are a 'free-er' country and a 'more democratic' country rather than simply being 'free' or 'democratic'. Yet these latter unqualified descriptions are frequently used, and perhaps even uncontroversially so if we remember that they are just handy labels for distinguishing the Western nations from other kinds of political system. They are convenient labels because they serve to suggest *the main points of comparison* with other forms of political organisation: we arguably have greater freedom of choice and greater powers of self-determination. However, when the descriptions are used to suggest more than important points of contrast, when they are used in an absolute sense and accepted as such, then we have a dangerous situation: we have a complacent populace which is prepared to believe it has actually reached the end of its long historical struggle for liberation and empowerment. Secure in the knowledge that they live in a 'free' country and a 'democracy', the mass of the people are ready to submit to any government which can convince them of its respect for freedom and its democratic credentials.

152

In turn, this political insouciance leads to intellectual complacency, to the feeling that one's belief system is more or less perfect and complete. In the modern Western world we seem to have reached this point, at least with respect to our beliefs about the proper organisation of community life, having finally hit upon the 'ideal' form of political organisation (the liberal democracy) coupled with the 'ideal' arrangements for production and distribution (the market economy). Such a complacency will not be easily shifted even when it is confronted with the most serious anomalies, and, in any case, if the anomalies are acknowledged as such, there will always be the last line in the capitalist's defence: an (often justifiably) unfavourable comparison of other political and economic régimes with our own and an implicit suggestion that if we undermine the status quo too much – if we search for something *better* – there will always lurk the threat of something worse.

I have identified freedom as the key concept in understanding both the appeal and relative success of capitalist society, but we need to look more closely now at the political structure referred to above – the modern liberal democracy – since this organisational structure is often thought to be the ultimate embodiment of the people's freedom to choose, perhaps even the greatest achievement of the Western World.

Under the 'democratic' form of political organisation, the electorate, is able to indicate by voting which values it wishes to see embodied in government policy; the party which most closely reflects the values of the majority who express their preferences through the voting procedure is thereby elected to power. In addition, the party that has gained power is wise if it is seen in practice to adhere to the values which it initially claims to represent, otherwise it risks losing popular support and, ultimately, political power. In its ideal form, liberal democracy provides the means for the general public to determine the politically effective value-system, and public opinion also remains a key variable in the working life of all democratically elected leaders.

However, reality departs from the ideal. Public opinion, for instance, may be rather pliable and may be moulded by governments in such a way as to support their continuation in power.[6] However, such criticisms will not concern me here as my current interest is in tracing the links between different strands of an ideology of which the *ideal* conception of liberal democracy is a part.

What I have outlined so far are the defining characteristics of a political democracy. There are many nation-states calling themselves political democracies and they differ widely in their constitutional particulars. What they all claim however is that the chosen goals of the government are regulated in some manner by the expressed preferences of the populace; if the populace values low inflation over full employment then the elected government will supposedly reflect this valuation in its policies. But there is another important aspect to the democracies of the West: traditionally, governments are not only supposed to be sensitive to the values of the electoral majority who bring them to power, but to the values of the many minorities who also lie within their jurisdiction. Whatever the other values that a government is charged to promote through its actions, in a true liberal democracy the value of toleration, permitting diversity in beliefs, is of primary importance.

This liberal aspect of modern Western political democracies assures relative continuity and political stability. Where diverse beliefs are tolerated instead of repressed there will naturally be less need for sections of the populace to rebel in order to force recognition of their beliefs and practices. In addition to toleration, the democratic structure of government holds out the possibility in principle that any one of the many tolerated value systems may come to influence government policy. So long as people's practices do not interfere with the rights of others, protected under the law, to live as they too see fit, they can be accommodated within such a society. This capacity to be so accommodating is undoubtedly one of strong points of the liberal democratic structure. The system is flexible and appeasing, whereas the less accommodating political alternatives remain comparatively brittle and risk being smashed under the pressure of stifled public opinion.

The modern liberal democratic structure, in contrast to classical liberalism,[7] presupposes the following: there is no way of *rationally* determining which set of values should predominate in our society. That is to say, there is nothing which it is objectively right to value, there are only those things which people do, as a matter of fact, value and we may discover what these things are by simply asking people for their opinions. This is what effectively happens in a democratic election: we do not necessarily get the *best* of the possible governments, we only get the government that people have chosen. Thus, the result of an election will, at best, simply reflect public opinion and in no way constitutes a rational argument for accepting the majority's value system for oneself. Even in a liberal democracy where

154

there are many distinct value systems allowed to exist side by side, one is expected nevertheless to defer on at least some occasions to the dominant value system, for if this were not the case, the government would have no effective authority. Those whose values it represents would happily act in accordance with any dictates conditioned by those values, whilst those whose values differ from the majority would simply go their own way. Effective political authority requires that the individual accepts as decisive the rulings of the government even when the individual is not convinced of the wisdom or morality of particular rulings. Thus, it can be seen that the main difference between a liberal democracy and other forms of political organisation is that its leaders will not attempt to legislate over as many of the beliefs and practices of its citizens as do the leaders of other, less accommodating, political régimes. It is not the case that governments do not insist on conformity in a liberal democracy, but rather that our leaders keep their demands to conform to an absolute minimum consistent with maintaining their effective authority and the identity of the nation-state. It is thus true that the governments of the West tolerate a greater diversity of opinion *to be expressed* but when it comes to a conflict between individual opinion, its resultant actions, and the law, then their tolerance must surely come to an end, just as it must in *any* nation-state.

There are those, like Honderich[8] and Singer,[9] who argue that civil disobedience has a place within liberal democracy. Neither has gone so far, however, as to contradict the view that we do *most of the time* have a duty to obey the law, that the government, in general, has the right to command our obedience.

A limited amount of civil disobedience serves to enhance the image of liberal democracy as a system tolerant of divergent opinions – even when these opinions are not being expressed through the 'proper' channels. Those who argue that civil disobedience has a place within liberal democracy recognise the role that it has in strengthening the structure as a whole, enhancing its flexibility and, by making concessions, ensuring that the greater part of the government's authority endures.

Clearly, the entire debate about civil disobedience only makes sense as a debate posing a moral dilemma if one is assuming that we do, ordinarily, have a duty to obey the law. But can we really have such a duty when, as Robert Wolff argues (*In Defence of Anarchism*[10]), it seems incompatible with our status as rational, autonomous agents to submit to the judgments of those who claim political authority? If we cannot *rationally* consent to be governed then the consent that is given is worthless: no-one can be morally bound by it any more than an adult can be bound by commitments made as an ill-informed and easily led child. Yet how can our consent to be governed be made consistent with our remaining rational and autonomous agents? This is, without doubt, the first and most formidable question of political philosophy and I find it by no means obvious that an adequate answer has ever been, or is ever likely to be, given. For now, however, I shall content myself with simply pointing out that a resolution to this debate (in favour of the authoritarian rather than the libertarian presupposition) has so far been assumed and, as argued in Chapter 6, has influenced the very ways in which we have chosen to conceive of the social scientific method. My concern now will be with the question of how a liberal democracy seeks to resolve the authority/autonomy conflict that Wolff has brought to the fore.

In a democracy we supposedly consent to be governed by those who represent the values of the majority. But is it rational to consent in advance to whatever may turn out to be the majority view, given that this may not also be one's own rationally determined view? Consent to be governed could only be rational if we presuppose that to be rational *means* to act in accordance with the majority. Thus it seems that in order to resolve the apparent conflict between authority and autonomous rational agency we must presuppose that our values derive what objectivity they have from the community so that what is really valuable or right or just in a given community becomes identical with whatever is predominantly believed to be right, valuable or just. In this way, the afore-mentioned conflict may conceivably be resolved and it turns out, conveniently, that we do indeed determine what is right or best by simply consulting the public.

But is this really satisfactory? We do not, it seems, ordinarily decide what is right on an issue by counting the numbers for and against certain proposals. We would ask for reasons, and not for numbers, in support of each position and in the end we would ideally be swayed by the best argument whether or not this is also the argument of the majority. Only at the national level, when considering government policy, do we begin to think in the way described above, allowing ourselves (even when we count ourselves as moral cognitivists) to be swayed by the force of numbers rather than by the force of reason. Why does this transformation in moral thinking occur?

In a *liberal* democracy, in exchange for the moral cognitivist's allegiance on any issues which s/he would be rationally bound to disagree, the government offers two things: (a) its toleration of that disagreement, and (b) its permission for that disagreement, together with the reasons for it, to be publicly voiced (freedom of speech) thus generating the possibility that the individual's views may eventually become the majority's. The moral cognitivist is usually pacified by these concessions – which are generous by comparison with all previous systems of government – and, in the meantime, is content to do no more than express his/her opinion and look to the future when, via the proper democratic channels, that opinion may have some impact on political decisions. The tolerance embodied in this system, together with the tantalising possibility of change for those who seek it, helps to keep the would-be 'trouble-makers' at bay with a level of success that no other political system has been able to achieve. Even when there is no realistic possibility of changing government policy, moral cognitivists who oppose some aspect of that policy will often do no more than to express their disapproval, being convinced that it is *objectively right* to do no more than to speak out about things that s/he thinks are *objectively wrong*. In short, there prevails the belief within liberal democracy that it is objectively right not to *do* anything about things which seem to be objectively wrong but which are nevertheless sanctioned and protected by the state. Both reason and conscience, which might otherwise goad us into action, are quietened by the fact that our reasoned opposition is given some outlet in this kind of society, guaranteed under the principle of Freedom of Speech.[11]

This brings us to one of the more fundamental elements of the capitalist belief system, that liberal democracy provides not only the best means, but the *only* means, for representing interests and resolving disputes in society. Any kind of direct intervention preventing what one perceives to be immoral (though legal) behaviour is often thought to lie outside of the boundaries of what is acceptable in a liberal democracy. Thus, although there *are* those who think civil disobedience has an important role to play in democratic societies, they often find themselves exposed to the criticism that illegal protest is undemocratic; would-be protesters have lost their case playing by 'fair' means so they have elected to play by means more foul. The ground rules are: we may *say* we do not agree with something we see happening, and we can try to prevent it through the established legal procedures, but if that method does not work for us we are ultimately obliged to tolerate the perceived wrong-doing just as others are obliged to tolerate our opinions in condemning it. So, at the end of the day, the morally-motivated campaigner who otherwise accepts the liberal democratic framework simply looks on passively at what s/he takes to be an affront to justice and decency, and will even take his/her passivity to be virtuous: such a response demonstrates the individual's tolerance and obedience to law. What is interesting, however, is that the response of the law-abiding moral objectivist does not differ in substance from that of the thorough-going *subjectivist* also working within the liberal democratic framework. They both end up doing (or not doing) the same thing, and the would-be threat to political authority posed by the *objectivist's* moral condemnation is effectively defused. Whatever threat remains, posed by those who uphold the case for change through civil disobedience, is kept to a minimum by the proponents of the case themselves, for they do not typically deny that the state has a right to command *most* of the time.

It is easy to see why Nietzsche attacked the morality of modern Europe, calling it the 'last man' morality and the morality which signals the triumph of the 'mentality of the herd'. To effectively give way to the majority when the majority is seen to embrace folly is to cease to exercise the will, or at least to no longer stand by the judgments of the will and to strive against whatever contradicts it. As I have argued in Chapters 3 and 5, Nietzsche rightly judges a morality which demands no positive action from its adherents to be weak and sickly, indeed to be almost no morality at all.

Whilst I sympathise with Nietzsche in his condemnation of many of the features of modern society, I do not recommend the Nietzschean response to what I have taken to be a problem of alienation. It is not *domination* (will-to-power) which overcomes the separation of the subject from all other objects but rather, the act of *identification*, which is the imaginative act, making your interests mine and *the* world *my* world.

Although the profoundly relativistic *modern* variant of liberalism is philosophically at odds with moral objectivism, it succeeds by the means described above in rendering it impotent (in contrast to the sort of moral objectivism espoused by Wolff), and indeed it must do so, because any moral objectivism that demands more than the mere expression of beliefs, or at best the selective flouting of the law, must surely pose a threat to political authority. Since 'liberal democracy' refers to a way of organising the state, of legitimising political authority by claiming that it accords with the will of the majority, it follows that liberal democratic forms of government must also have the means for dealing specifically with moral objectivist dissenters. In all forms of social and political organisation we have seen the ruling élite seek to enforce on the majority a value system which has been taken to be objective; insofar as they succeed (by education or other means) in extending the number of adherents to their value system to that extent their political authority was reinforced. However, there are always moral objectivists who have conflicting values and these, being repressed, would remain to undermine political authority. Only in the modern Western state is there no longer any need to uphold the authority of governments by promoting a *single*, supposedly objective, value system until one has obtained uniformity in belief. In an important sense, all value systems are now treated equally and, in principle at least, a democratic government might reflect any one of them. At the heart of this transformation to an overtly more tolerant society is the belief, which has emerged as dominant, that no set of value judgments can ever be deemed 'objective'; they are all equally arbitrary and therefore there can be no reason to demand conformity to any one set of values. Anyone who thinks there is something *objectively* wrong with society can usually be dismissed as arrogant, and particularly so if they are in the minority.

Ultimately, alternate value systems are tolerated in a liberal society, not out of a sense of the objective value of toleration, but because the choice between sets of values is taken to be non-rational. It follows from the belief in the arbitrary nature of value choice that one may as well not interfere with other people's choices on questions which concern their values. It also follows that the politically effective value system may as well be the one that reflects the values of the most people for this will promote greater national stability.

What has effectively happened under modern liberalism is that we have achieved a more tolerant society, a society where we have more power to choose, but where any choice we make (regarding personal goals) has become intrinsically less valuable to us. We are free to choose but our choices are felt to be merely arbitrary. We are apparently free, but in our freedom we lack direction and purpose and so are deprived of the satisfaction to be found in the full exercise of our rational faculties.

A belief in the essential subjectivity of values is central to the whole structure of capitalist society. We have seen how this belief underpins the organisation of the political life of our society, and it is but a short step to see how it conditions our economic life. The current dominant paradigm in Western economics is one which assumes a subjective theory of value. An *objective* theory of economic value, based on the quantity of labour crystallised in the goods produced, is associated with Marx and with those

159

writing in the Marxist tradition, and as such it is often linked with attempts at critiquing the capitalist system. The most effective way to off-set criticism coming from this sector is to attack the theory of value which underpins their case. Hence it was the neoclassical economists – following the lead of writers like Menger,[12] Jevons[13] and Walras[14] – who finally replaced the labour theory of value with a theory based on subjective satisfactions, or 'utility'. Clearly a person might derive greater utility from a product than her next-door neighbour so that its value *to that person* is greater. However, the product's final value on the neoclassical view depends not just on what one person is prepared to pay but on what consumers in general are prepared to pay (effective aggregate demand) and, of course, it will also depend on the price the producer is prepared to accept in order to supply it. In the end, the true value of the product is supposedly reflected in its market price, which is determined by the forces of supply and demand. 'Market forces', therefore, are basically just the forces which emerge from individuals' (as producers or consumers) subjective valuations of a wide variety of goods and services.

With utility, and the forces it gives rise to, at the heart of our economic life, the outcome of our economic activity will be a distribution and use of wealth which reflects the amalgamation of individuals' subjective, *non-rational* preferences. Sometimes it may happen that society (or the majority and its political representatives) does not like the outcome arrived at via the unfettered operation of market forces. In such cases, there might well be some government intervention, though with a basically 'free market' economy, intervention will necessarily be against the norm.

The parallels between a more or less 'free' market model of economic life and the liberal democratic model of political life are striking but, of course, not coincidental. The outcome of economic decisions based on derived utility is an outcome which, ideally, best satisfies the subjective preferences of the majority of consumers, just as in the political sphere, the outcome of an election is a government which best satisfies the majority's *political* preferences. In both sectors, it is the expression, through the acts of purchasing and voting, of people's non-rational, value-based preferences, which 'buys' for the public the outcomes it desires and there is in neither case anything to guarantee that what the public has 'bought' is either objectively good or even the best of the available alternatives. In fact, there is the tacit assumption in modern capitalist society that what is chosen either by the public for the public, or by the individual for the individual, is *necessarily* good, because there is no more to being 'good' or 'the best' than being what is actually desired and freely chosen. The possibility of making an alienated choice, that is, of having a value-based preference for something which is objectively bad, is not admitted.

At an earlier point in this chapter, I stated that we would be looking first at the main characteristics of societies where the capitalist ideology predominates in order to determine the most fundamental beliefs of this ideology. We have reviewed the nature of the modern liberal democracy, its connection with the free market economic system, and, briefly, the role (and nature) of society's commitment to the principles of toleration and freedom of speech. We have not yet traced the connections between any of the above-mentioned characteristics and the most commonly-held conception of science as value-neutral and therefore objective. It would be useful now to finally spell out the connections between science and the ideology-type under examination.

Science is seen as the exemplary case of a rational human enquiry, widely perceived to be straight-forwardly concerned with matters of fact, and for that reason, is thought to provide determinate, objective answers to our questions. This state of affairs, wherein science is held in high public esteem, represents the impact upon society of a philosophical idea, namely, an epistemological divide separating 'facts' from the more controversial realm of 'values'.

In Chapters 1 and 2, I have examined the impact on human knowledge and the human psyche of what I have called an *alienating* conception of rationality. It is a conception which proves destructive in two respects. Firstly, it undermines our confidence in an independent reality by setting us at an epistemological distance from what is to be known, and secondly, it undermines our confidence in ourselves by denying our rational capacity for embracing (or for identifying with) what is real. In attempting to define what rationality is, we are necessarily laying down the limits of what we expect a rational being to be able to achieve. If we define rationality in a *more* limiting way, we are expecting rational beings to achieve *less* even in the full exercise of their rational faculties. Yet however we decide to define rationality, we must surely define it in such a way that something remains which is rationally possible. In other words, we cannot draw in the limits of what the rational individual may achieve so completely as to render the set of rational decisions an empty one. If our conception of rationality leaves us with nothing that can be rational then we must abandon it for a better one.

The classification of forms of enquiry into two utterly distinct categories, one type concerned with fact and the other with values, is rooted in the overly limiting and ultimately self-destructive conception of rationality which was examined in the earlier chapters. Although the distinction is clearly meaningful, it has gradually come to epitomise a certain way of thinking about the world (and our place within it) which is at once harmful and misleading. More specifically, the dichotomy[15] has come to be associated with a conception of objectivity, based on fact and excluding values, where the facts are seen as coming from the object, and values are seen as an unwelcome intrusion coming from the subject and for that reason to be avoided at all cost. Thus, the subject apparently gets in the way of his/her own efforts to acquire objective knowledge. If it is subsequently discovered that the subject cannot 'get out of the way' then all hopes of attaining objective knowledge will be dashed.

It is the above conception of objectivity, conditioned by the fact/value divide, which governs popular thought about science in that science (or at least good science) is usually perceived as being concerned with the facts, examining them in a characteristically value-neutral way. In addition, it is this conception of objectivity which has influenced our ways of thinking about society and it is, as we have seen, closely related *through the implied subjectivity of values* to the economic and political structures which are currently in place in the Western world.

That there are intimate links between this conception of science and the socio-economic order referred to as capitalism, is, in the first instance, strongly implied by the historical coincidence of the eras of capitalism and of rapid scientific progress. On the one hand, capitalism gave the impetus to scientific advancement as the demands of competition in industry boosted the need for greater technological inventiveness. On the other hand, science supplied the capitalist belief system with an apparently firm epistemological bedrock: scientists were to be the new experts of the era, replacing the clerics of previous times, and replacing their supposed myth and superstition with *real* knowledge based on value-free observation. In the first stages of capitalism, non-scientific beliefs (for instance, moral or religious convictions) coexisted with science, but such a coexistence was bound to be uncomfortable – though not especially for the scientist. Inevitably, it was value-based assertions which were to come off badly in a society where particular meanings for words like 'objectivity' and 'rationality' were being promoted, if not shaped by, an ideal and widely popular conception of science. The conception of scientific objectivity as essentially value-neutral naturally posed a threat to the plausibility of claims concerning a possible *moral* objectivity. For value choices to be objective, their objectivity would have to be of a radically different sort to that claimed for science.

Classical liberals such as Mill believe that some basis can be given for making rational choices in ethics. In Chapter 5, I compared the basis for morality provided by Mill's utilitarianism, with the kind of foundations for morality provided by philosophers under the influence of *modern* liberalism. The contrast is striking, as the shift from classical to modern liberalism marks the further retreat of the subject into his/her own sphere. Mill's approach demands a going-beyond-oneself to identify with the interests and pains of others and to count them equally with one's own; freedom of choice and tolerance are *positively* evaluated in a way which they are not subsequently under modern liberal systems of thought. In a system of thought dominated by a subjective theory of value, the goal of increased toleration and of protecting individual freedom of speech could not consistently be claimed as justified because of its *objective* goodness. Rather, these essential liberal goals are taken on board for the *negative* reason that we lack any value-based justification for interfering with the arbitrary, non-rational choices of others. Typically, modern liberal thinkers who have wished to secure the basic 'moral' features of liberal, free market economies have had to resort to the use of 'hypothetical world' devices, like Rawls's 'Original Position', which ensure that we make decisions about the *real* world with absolute impartiality. The oddity of this approach lies in the fact that it attempts to derive principles of justice – *moral* principles, which concern how we may treat others – without requiring that we identify others as individuals, or that we identify *with* them. In effect, the circumstances described in the 'Original Position' require no more of us than self-interested thought which is converted, when the 'veil of ignorance' is lifted, into *impartial* principles of justice. What such devices produce is a morality without requiring that we know of, or sympathise with, the beneficiaries of our apparently 'moral' behaviour.

In comparison with the morality of classical liberalism, as represented by Mill, the morality of the modern liberals appears as a sham. It does not demand that we get involved with others or experience their interests as of equal weight to our own. What it offers instead are some self-interested, or prudential, motives for granting others certain minimal conditions of life, the same conditions which we desire for ourselves. Thus, what morality amounts to is an exchange: I'll observe your basic rights if you observe mine; I will observe your rights *in order* that you will observe mine.

Throughout this work, I have chosen to characterise alienation as, essentially, a form of isolation provoking various irrational responses. We may feel isolated from other people, from nature or from our natural impulses, or, more fundamentally (when our confidence in our ability to directly know the world is undermined) from reality as such. I have also suggested in earlier chapters that our response to this problem of isolation can take two forms: we can withdraw further and deny the reality of what we are isolated from, or else we can seek to close the gap between ourselves and the 'other' by means of domination, possession, or control. The morality of modern liberalism described above represents the first response, the complete withdrawal of our involvement with others; other features of our society – the endless striving for economic growth, rampant consumerism, the evident drive to dominate and control nature, to 'harness' its forces for ourselves, the maintenance and further extension of personal power in business, politics, and even private life – all of these are aggressive responses which, given the over-assertion of the subject, tend only to succeed in obliterating the object. Neither of these types of response will ultimately overcome the experience of alienation.

In retaining a commitment to objectivism in morality, classical liberalism kept the earlier form of capitalist ideology at least a step away from the more intensely alienated – because morally impoverished – ideology of advanced capitalism. However, whilst differing in their approach to ethics, the classical liberals did not substantially differ from later liberals in their conception of science. The science-dominated conception of objectivity was not challenged, and thus it remained to undermine the epistemological status of moral judgments. It was the failure to challenge the existing value-neutral conception of scientific objectivity which paved the way for the mutation of classical liberalism into its modern, subjectivist and relativistic variants.

There is a tendency, which I remarked upon in Chapter 7, for ideologies to become more internally consistent in response to criticism or a threat to their continued influence. This is what we have seen in connection with the capitalist ideology. Modern, relativistic liberalism sits better with the other component parts of the capitalist belief system than did the moral objectivism of the classical liberals. Aside from the greater internal cohesion achieved by the transition to modern liberalism, there is a further gain: the ideology of the West now lays an even stronger claim to some of the most powerful and appealing concepts. It is the ideology of a social order which can, with ease, identify itself with such positive notions as democracy, freedom, toleration and diversity of choice and has been able to use these most appealing concepts to achieve unprecedented levels of acceptance for a system of government and for a mode of economic life. Any form of moral objectivism, by contrast, appears in our society as an arrogant intolerance of alternative views. Marx, understandably, could not have foreseen the tenacity of the capitalist economic system since he was unaware of the developments in the belief system which were to take place. Perhaps more than any other factor, it is the all-accommodating relativism of modern capitalism which protects the system from the revolutionary forces it provokes.

There is an additional, significant way in which Marx is mistaken about the nature of capitalist ideology: he never identifies the ideal of objective, value-neutral science as an integral part of the ideology he is attacking. In neglecting this point, he misses an important clue to understanding the nature and extent of human alienation. In Chapter 6, I have attempted to show how a particular conception of scientific methodology (in both its simplistic 'means-end' and more sophisticated 'programme-prognosis' forms) is conditioned by unexamined value-judgments concerning the social order. In a society like our own (ie. a hierarchical one) it is important for science to be perceived as impartially pursuing the facts because at the end of the supposedly value-neutral investigation the information gained will be needed to bestow intellectual authority upon those who may wish to make use of it in accordance with their own value-related aims. As Myrdal recognises, the claim of the social scientist to be authoritative, to have, therefore, a 'legitimate' influence on government policy, depends on sustaining the non-perspectival, non-value-based, view of social scientific objectivity. It is only by retaining a hard core of objectivity in the non-perspectival sense (knowledge without a knowing subject) that knowledge thus acquired can be used on behalf of all rational subjects to significantly shape their lives *without*, apparently, undermining their status as rational, autonomous agents. What is important then, in a hierarchical society, is that we have a methodology for science which reflects the need to separate the thing known from its subsequent use; science must simply discover the means, or provide the 'prognosis', whilst society and its democratic representatives determine the end (or the content of the 'programme') to which this knowledge will be put. Thus the scientist is absolved of all

165

responsibility for the use made of his or her work, and the public is relieved of the most awesome task of trying to understand the world for itself. The public need simply state its requirements, through its representatives, and science will provide the means for their achievement, apparently without the rationality of either group being in any way compromised by this 'division of labour'.

Marx argues[16] that the science of classical economics is not a science of eternal truths, like physics or chemistry, but serves only to explain the economic interactions within a particular kind of social order, mistaking their specificity for generality. Should the social order be transformed, the 'laws' of economic interaction would change and classical economics would become superfluous. Marx, however, does not fully envisage that our conception of scientific methodology might itself be subject to transformation. Our views on what is ideal in scientific method – for instance, the striving for value-neutrality in the belief that this guarantees objectivity – are in their turn conditioned by the prevailing social order and by the position that is allocated to the rational, would-be self-determining agent within this order. The lowly position of the enquiring subject, who, to achieve objectivity, must 'get out of the way' is reflected in current conceptions of scientific method *and* in the society which holds this conception in high esteem.

I said above that the adoption of a value-neutral conception of science, based on up-holding some form of means-end dichotomy, runs a risk – if it is found that value judgments cannot be disentangled from the scientific 'facts', or that means cannot be disentangled from complex sets of ends, we must give up hope of attaining objective scientific knowledge. In the last few decades in the positivist philosophy of science we have witnessed just this retreat from the possibility of an objective science. In the light of contributions from philosophers like Kuhn and Feyerabend it no longer seems possible to demonstrate how scientists make important choices about theory development, to demonstrate how decisions are made between competing research traditions or paradigms, *without* admitting the role of value judgments in these decisions. Kuhn and Feyerabend, among others, express the fast-spreading realisation that science is *not*, and never has been, straight-forwardly concerned with the 'fact' side of the great fact/value divide. Thus, in the second part of Chapter 6, we looked at the problem of rational choice between theories *given* the relativist's assumption that a subjective contribution to the choice cannot be eliminated.

166

To summarise: the fact/value gap, which has dominated our thinking for so long and in so many ways, began (historically speaking) by wreaking havoc with our conception of the status of moral thinking, leading ultimately to moral subjectivism and relativism which are expressed in common place fashion by so many people today. In the second instance, and at an historically later date, the emphasis on the fact/value divide began to weaken that part of human enquiry which had once been thought to be unquestionably rational and unambiguously concerned with fact, namely science. Thus we see that a certain, doubtful construal of the relationship between our knowledge of facts and values has apparently undermined – or at least has begun to undermine – all major areas of human knowledge. There is nowhere left to turn for the epistemological security we appear to crave.

What I have been suggesting as a way out of the impasse in which we now find ourselves is a fundamental revision of the fact/value relationship, and indeed a revision of the way in which we view the process of knowledge acquisition generally. Perhaps the most important change we must make is in the way we view the role of the enquiring subject.

Throughout this work I have been using the term 'interpretation' in a realist way, a way which refers to the action of coming to know for oneself the way things really are. This is a usage which can be contrasted, rather than simply confounded, with 'interpretation' understood as a form of meaning distortion, (ie. it is contrasted with 're-interpretation' – if reality is re-interpreted it is distorted, not understood.) Given this proviso, it follows that a good interpretative framework for understanding and generating explanations of the world is one which allows us to grasp what the world is really like, albeit from the limited perspective of the interpreter. I would argue that the role of the interpreter has for too long been misunderstood because our sound intuitions regarding what is required for objectivity have been misdirected: it is the interpretative framework which must be transparent, which must be prevented from 'getting in the way', and not the enquiring subject, or interpreter, who must remain inactive and self-effacing if knowledge is to be at all possible. Knowledge acquisition is a type of action and as such it essentially requires an active, reality-interpreting subject.

In Chapter 7, I have addressed what I see as the real problems threatening scientific objectivity and progress. Bias is the lesser, more easily remedied, failing on the part of individual scientists. The failure to recognise the over-

167

arching role of a wider, possibly ideological, belief system is a far greater failing not as readily acknowledged. It is not the incursion of value judgments as such into our body of theoretical knowledge which is worrying, but the incursion of unexamined, and therefore undefended, valuations. The existence of such valuations means that the interpretative framework may not achieve the transparency referred to above, for, without being subject to critical examination, the interpretative framework becomes inflexible and our capacity to explain phenomena and to understand the world becomes correspondingly limited. The assumption that science is, for the most part, value-free has been damaging because it has restricted debate, directing attention away from the subject's stance towards the world s/he is seeking to understand, and away from the framework for interpretation which reflects this stance.

All theoretical knowledge has a subjective foundation, at least in the sense of its belonging to a knowing subject. Theoretical explanation is only possible because of (a) the existence of an enquiring subject, wanting explanations of the various phenomena, and because of (b) the existence of an objective, independent yet knowable world, which is the proper object of our enquiry. To change one's view of the world (or one's theory allegiance), arguments must concern (a) features of the world (the facts) and/or (b) our relation to those facts, (that is, our stance or perspective which may be conditioned by an ideology). If only (a) is addressed then certain explanatory alternatives will never come to seem plausible. An alteration in the 'lived relation' to one's environment may be necessary in order to enhance one's understanding of reality.

2. Some Tentative Recommendations

It seems to me that the enquiring subject can no longer be viewed as *necessarily* a hindrance to itself in its own efforts to acquire objective knowledge. True knowledge requires the subject's involvement with what is real, and not, as seems to have been believed, the subject's impassive indifference. Certainly, impartiality would seem to be a requirement for objectivity but even this demand can be understood in more than one way: I take it to mean that it is the selectivity in our *involvement* with reality which threatens objectivity, rather than our selectivity in becoming *disengaged* from it. Thus, to open up the range of explanatory possibilities, to push debate forward where it has become stagnant and confrontational, I recommend a lifting of restrictions formerly placed on scientific debate. Firstly, scientific objectivity does not require a 'standing back' attitude so much as a keenness to get involved, to immerse oneself in the object of one's study. (Perhaps this is what all our really great scientists have always done.) In line with this, it seems that – contrary to popular belief – arguments designed to stimulate imaginative involvement cannot be regarded as absolutely illegitimate. Indeed, in the social sciences particularly, such arguments may have an important role to play in raising awareness of aspects of social or economic reality which are not directly experienced by the researchers themselves. In other words, arguments which stimulate an under-active imagination and engage the sympathies may correct a tendency towards selective involvement (e.g. with groups sharing one's economic or class interests). In this manner they may contribute to greater objectivity and enhance the possibilities for progress by breaking down the otherwise insurmountable barriers to mutual comprehension.

The second, and perhaps more far-reaching, recommendation concerns the value-freedom doctrine. In both the natural and the social sciences there have been problems explaining how theory selection and development may proceed rationally. The problem has centred on the apparently ineliminable role for human judgment in deciding how theories might profitably be developed and which theories, or theoretical traditions, are to be preferred. Since it had been assumed that conceding a decisive role to the enquiring subject undermined objectivity, the rationality of the scientific enterprise now appears under threat. The demand for value-neutrality, however, arises only on the assumption that value-relatedness (ie. a value-based input from the enquiring subject) runs absolutely contrary to claims of objectivity for any knowledge acquired and it is this assumption which may now – and indeed *should* now – be seriously challenged. The value-freedom doctrine, and related assumptions concerning the essential non-rationality of value choices, has to be abandoned.

If our conception of rationality leaves us with nothing that can be rational then the time has come to review what we mean by rationality. If science cannot be a rational enterprise through being value-free then we have to ask: can it be rational *granted* that we acknowledge the role of subjective valuations? I have tried to argue that it is only certain kinds of input from the enquiring subject which are detrimental to our efforts to obtain objective knowledge. Valuations, which are an integral part of the interpretative framework, only undermine our ability to grasp reality when they remain unexamined and assuming their essential non-rationality is one excuse not to examine them; the interpretative framework then becomes inflexible and our chances of finding the best, the most all-embracing explanation of phenomena, are reduced. Thus, maximum flexibility and an awareness of one's own fallibility are the key elements required for scientific progress. There is nothing *certain* at the base of the scientific edifice, nothing that actually has the security once thought to belong to facts, but recognising the fallibility of the process of theory development does not in any way undermine the claim that theories tentatively developed may nevertheless be objectively true.

In this final chapter I have tried to draw out the links between the value-neutral conception of science, a certain way of organising society, and a belief system which is ideological in that it alienates the individual from important aspects of reality, not least, from *other* individuals. It is a belief system which leaves humanity without a sense of purpose or direction because an integral part of that system is the assumption that value choices are non-rational. In an important sense therefore it is thought not to matter what we do or what we choose to value; all that matters is that we have *freely* chosen. And yet what satisfaction is there in exercising freedom when there is apparently nothing *really* worth choosing? This, I think, is one of the primary causes of our modern-day malaise. To make matters worse, even confidence in the objectivity of science is on the decline. A populace that no longer thinks it matters what we believe, even in what was formerly the realm of fact, is one which is all too ready to be manipulated in Orwellian fashion by those with the power and inclination to take up this opportunity. Without the touchstone of reality we are simply washed along with the majority, caught up in convention without the critical resources to muster any reasoned opposition.

The doctrine of value-neutrality in science, together with the value-impoverishment of modern liberal society, have immunised us to criticisms which might otherwise undermine them. The valuations enmeshed within our framework for interpreting experience have become frozen in the case of supposedly value-neutral science, or else float freely *above* the reach of rational criticism, as is the case in a profoundly relativistic, liberal society. The result for society as a whole is a weakened basis for moral and intellectual opposition to any ideas, theories, or ways of life.

What is needed is the restoration of the basis for rational evaluation and criticism of alternate world-views, theories and/or practices. In short, what we need is to find ourselves brought back in touch with reality, indeed to find that we need never have lost our grip: the real, independent, and ultimately knowable, world was always there waiting to be understood and all that was lost was the confidence in our own ability to, *by rational processes*, come to understand it. It is the undermining, or the under-estimation, of the powers of the enquiring subject that has led us to believe we can have no firm grip on the independently-existing object.

My final recommendation, then, concerns the individual and the appropriate attitude to have towards one's experiences. For too long the individual has felt him/herself alienated, pushed back and left at a distance from reality. Philosophy itself has told us this, but philosophers have not been alone in telling the layperson that s/he cannot really *know* what is real or true. In a hierarchical social order, the 'ordinary' individual defers to some expert or other on any matter of importance. It is only because of an interesting twist in history that, in the most recent stage of our social development, there is *no-one* who purports to be an expert on questions of value. But this does not mean that the layperson's view is now *as good as* anybody's. Certainly, there is no expert to defer to, but now the layperson's view is only worth *as little as* the next person's. This being so, s/he must defer to the majority or risk being labelled arrogant or intolerant.

If, however, we take a fuller, more constructive, view of human rationality (one that does not ultimately undermine all major areas of human knowledge) we see that the ordinary individual, the rational enquiring subject, must *as such* have greater powers than have thus far been supposed. To possess rationality means to have the capacity to embrace reality fully, to make the real world identical with the world of one's subjective awareness. Only in exercising this capacity do we realise our full potential and eliminate the gaps separating subject from object, separating one individual from another, and thus genuinely alleviate the problem of alienation.

The rational subject, the ordinary individual, must now be elevated to his/her proper place. The layperson has always been prepared to mistrust his or her own judgments, to be cynical as regards his or her own capabilities, and yet naïve as regards the judgment of experts or leaders who, apparently, have no similar misgivings. The individual's viewpoint is not a perspective from which nothing can be properly seen in contrast to the position of the expert or elected leader; it is the *only* perspective from which anything is ever known; there is nothing that is not known (if it is known at all) to an individual subject. This being so, each individual can only rationally be his/her own final authority on questions of truth and propriety since no subject is in any more privileged a position to judge than any other. As rational beings, therefore, we must settle questions, however provisionally, for ourselves instead of entrusting their settlement to others on our behalf. What this entails, then, is a reversal of the attitudes mentioned above: our cynicism should be saved for those who seek to act on our behalf, without our accepting or perhaps even comprehending their judgments. Our trust, on the other hand, is best placed in our own capacities to understand the world, in the reality of the objects and people that stand before us but who need not *remain* at a distance if only we would choose to exercise our rational capacities and, through knowing them as they really are, bring them into our personal sphere.

Notes

1. I use the term 'modern liberal' to contrast with the classical, or Millean, form of liberalism which does *not* exclude the possibility of an objectivist moral philosophy.
2. That there *are* serious problems facing society is a point which is becoming increasingly difficult to deny. Many ordinary people were forced to consider the question of what is going wrong with society in the wake of the Dunblane killings. The suggestion that we need elaborate security systems in place in all our primary schools and nurseries only serves to bring out the insanity of modern society all the more plainly. Add to this the problem of homelessness, unemployment, bullying and high levels of stress in the work place, family disintegration, looming environmental disaster and, in the face of it all, a widespread addiction to gambling as people increasingly see the lottery ticket as their best (!) hope of salvation. – At what point are people going to notice that there is something *radically* wrong with society?
3. Countries of the old Soviet Union and Eastern Europe are perhaps examples of this – though disillusionment may swiftly follow for countries who are effectively the 'poor men' of the market place.
4. For instance, this would have been Godwin's position. He argues that the abolition of the state was the *pre-requisite* for free, responsible decision making. More recent examples of this position in political philosophy are E. Goldman (see *Anarchism and Other Essays* and RP. Wolff, *The Poverty of Liberalism.*)
5. Stevenson, CL., *Ethics and Language*, London, 1944. Stevenson discusses persuasive definition on p227.
6. In particular, there is the possibility that those with political and economic power employ the resources of the media to shape public opinion in ways favourable to their personal political interests. For instance, using the media to generate and sustain

feelings of nationalism and aggression towards foreigners has, in the past, increased the likelihood of a Conservative election victory.

7. Mitchell, B *Law, Morality and Religion*, Chapter 6, 'Varieties of Liberalism'. Also Lee, K., in *A New Basis for Moral Philosophy*, makes a distinction between the classical and modern variants of liberalism.

8. Honderich, T., *Violence for Equality*, Routledge, London, 1989.

9. Singer, P., *Applied Ethics*, Oxford University Press, 1986.

10. Wolff, RP., *In Defence of Anarchism*, Harper & Row, 1970.

11. It is worth emphasising that, when it comes to appraising the actions of governments, our moral sense seems significantly altered. Would we allow anyone to kill children, satisfying ourselves with a mere 'protest' at their actions? Many would recoil in horror at the suggestion of such moral cowardice. Yet was there anything more than a (popularly condemned) protest when the British government sent out the fleet to destroy (among others) 14 year old Argentinian conscripts? Of course *their* being there was the fault of the Argentinian government, not of *our* government; and of course you have to take into account the big political picture, the international ramifications of military inaction, & etc. In most cases, the rationalisation for such actions follows swiftly, but the fact remains: we judge the ordinary individual according to one set of moral intuitions and those acting on the authority of government by quite another. It seems that governments exist on a different moral plane, one which permits us to excuse all sorts of atrocities by raising them to the level of political necessity.

12. Menger, C., *Principles of Economics*.

13. Jevons, WS., *Theory of Political Economy*.

14. Walras, MEL., *Elements of Pure Economics*.

15. It has been argued that the fact/value distinction is more than a dichotomy even; it has been called a *dualism* (in Plumwood, *Feminism and the Mastery of Nature*). In any case, the distinction is so firmly rooted it has affected the way we view the discovery of facts. The process of coming to know something has been systematically misrepresented.

16. In the *Economic and Philosophic Manuscripts*, for example, Marx criticises classical economists for failing to view economic phenomena within the context of an historical, evolutionary process.

Bibliography

Apel, T.W. (1982), *Against Epistemology*, Basil Blackwell, Oxford.

Aristotle, *Metaphysics*.

Ayer, A.J., 'I think therefore I exist', in H. Morrick (ed.) (1981), *Introduction to Philosophy of Mind*, Dover Press.

Boyd, Gasper and Trout (eds) (1991), *The Philosophy of Science*, The MIT Press, Cambridge, Mass.

Camus, A., 'The Absurdity of Human Existence' in E. Klemke (ed) (1981), *The Meaning of Life*, Oxford University Press.

Comte A., 'Plan of the scientific operations necessary for reorganizing society' (Third Essay 1822), in G. Lenzer (ed.) (1997), *Auguste Comte and Positivism, the Essential Writings*, Harper & Row, New York.

Descartes, R., 'Discourse on the Method' in E. Anscombe and P.T. Geach (eds) (1954), *Descartes' Philosophical Writings*, Thomas Nelson and Sons, London.

Feyerabend, P.K. (1988), *Against Method*, Verso Books, London.

Feyerabend, P.K. (1978), *Science in a Free Society*, New Left Books, London.

Feyerabend, P.K.(1975), 'How to Defend Society against Science' in *Radical Philosophy*, vol.2.

Fromm, E. (1984), *The Fear of Freedom*, Ark Paperbacks, UK.

Gellner, E. (1974), *Legitimation of Belief*, Cambridge University Press.

Godwin, W. (1985), *Enquiry Concerning Political Justice*, Penguin Books Ltd., Harmondsworth.

Goldman, E. (1969), *Anarchism and Other Essays*, Dover Publications, New York.

Habermas, J. (1972), *Knowledge and Human Interests*, Heinemann Educational, London.

Harcourt, GC. (1972), *Controversies in the Cambridge Theory of Capital*, Cambridge University Press.

Hare, RM. (1963), *Freedom and Reason*, Oxford University Press.

Hayek, F.A. (1976), 'The Mirage of Social Justice' in vol.2 of *Law, Legislation and Liberty*, Routledge, London.

Honderich, T. (1989), *Violence for Equality*, Routledge, London.

Horkheimer, M. (1972), *Critical Theory*, Herder & Herder, New York.

Hume, D. (1978), *A Treatise of Human Nature*, Oxford Clarendon Press.

Husserl, E. (1970), *Crisis in European Sciences*, Northwestern University Press, Evanston, Illinois.

Jevons, W.S. (1957), *Theory of Political Economy*, Kelley & Millman, New York.

Jones, R.S. (1982), *Physics as Metaphor*, University of Minnesota Press, Minneapolis.

Kant, I., *Groundwork for the Metaphysic of Morals*, in H.J. Paton (1948), *The Moral Law*, Hutchinson University Library, London.

Keat, R. (1981), *The Politics of Social Theory*, Basil Blackwell, Oxford.

Koyré, A. (1968), *Metaphysics and Measurement*, Chapman and Hall, London.

Krige, J. (1980), *Science, Revolution and Discontinuity*, Harvester Press, Brighton.

Kuhn, T.S. (1970), *The Structure of Scientific Revolutions*, University of Chicago Press.

Lakatos and Musgrave (eds) (1970), *Criticism and the Growth of Knowledge*, Cambridge University Press.

Lee, K. (1985), *A New Basis for Moral Philosophy*, Routledge & Kegan Paul.

Leiss, W. (1972), *The Domination of Nature*, George Braziller, New York.

Lovibond, S. (1982), *Realism and Imagination in Ethics*, Oxford University Press.

Marx, K., *Economic and Philosophic Manuscripts*, in T.B. Bottomore (ed.) (1963), *Karl Marx: Early Writings*.

Menger, C., *Principles of Economics*, Dingwall and Hoselitz (trans.) (1950), Free Press, Glencoe, Illinois.

Mill, J.S. (1972), *On Liberty*, Everyman's Library, London.

Mill, J. S. (1979), *A System of Logic*, Longmans Green & Co., London.

Mirowski, P. (1981), *Against Mechanism*, Rowman & Littlefield, USA.

Mitchell, B. (1967), *Law, Morality and Religion*, Oxford University Press.

Myrdal, G. (1953), *Political Element in the Development of Economic Theory*, Routledge & Kegan Paul, London.

Myrdal, G. (1972), *Asian Drama*, Lane Publishers, London.

Myrdal, G. (1969), *Objectivity in Social Research*, Wesleyan University Press, Middletown, Conn.

Myrdal, G. (1958), *Value in Social Theory*, Harper & Bros, New York.

Nagel, T. (1978), *The Possibility of Altruism*, Princeton University Press.

Nagel, T. (1986), *The View From Nowhere*, Oxford University Press.

Newton-Smith, W. (1978), 'Under-determination of Theory by Data' in Supplementary Volume LII of the *Proceedings of the Aristotelean Society*, London.

Nietzsche, F. (1961), *Thus Spoke Zarathustra*, Harmondsworth, Penguin Books.

Nietzsche, F., *Unpublished Notes* in A.C. Danto (1981), *Nietzsche as Philosopher*.

Orwell, G. (1962), *The Road to Wigan Pier*, Penguin Books.

175

Parfit, D. (1984), *Reasons and Persons*, Oxford University Press.

Plato, *The Republic*.

Plumwood (1993), *Feminism and the Mastery of Nature*, Routledge, London.

Popper, K. (1972), *Objective Knowledge: an evolutionary approach*, Oxford, Clarendon Press.

Popper, K. (1966), *The Open Society and Its Enemies*, Routledge & Kegan Paul, London.

Rawls, J. (1972), *A Theory of Justice*, Oxford Clarendon Press.

Schopenhauer, A. (1969), *The World as Will and Representation*, Dover Publications, New York.

Singer, P. (1986), *Applied Ethics*, Oxford University Press.

Stevenson, C.L. (1944), *Ethics and Language*, London.

Walras, M.E.L. (1954), *Elements of Pure Economics*, William Jaffé (trans.), Irwin Press, Homewood, Illinois.

Weber, M. (1947), *The Theory of Social and Economic Organisation*, Wm. Hodge & Co. Ltd., London.

Weisskopf, W.A. (1973), *Alienation and Economics*, New York.

Wolff, P.P. (1971), *The Poverty of Liberalism*, Harper & Row, New York.

Wolff, P.P. (1970), *In Defense of Anarchism*, Harper & Row, New York.